# THE GEM & MINERAL COLLECTOR'S GUIDE TO IDAHO

Revised Edition

By Lanny R. Ream

Gem Guides Book Co.

315 Cloverleaf Dr., Suite F

Baldwin Park, CA 91706

Revised Edition 2000



Library of Congress Catalog Card Number 00-133087
ISBN 1-889786-13-6

Maps: Lanny R. Ream
Cover: Scott Roberts

NOTE:

Due to the possibility of personal error, typographical error, misinterpretation of information, and the many changes due to man or nature, *The Gem & Mineral Collector's Guide to Idaho,* its publisher, and all other persons directly or indirectly associated with this publication, assume no responsibility for accidents, injury, or any losses by individuals or groups using this publication. In rough terrain and hazardous areas, all persons are advised to be aware of possible changes due to man or nature that occur along the trails.

# Table of Contents

# INTRODUCTION

*The Gem & Mineral Collector's Guide to Idaho* has been available for several years in a two volume set. This revised edition combines the information of the two volume set into one volume. Some localities have been removed because they are now inaccessible, and new localities have been added. In addition, the addresses and information pages have been expanded to include all U.S. Forest Service and Bureau of Land Management offices. The two fee collecting areas, Emerald Creek and the Spencer Opal Mine, have been updated to include current pricing information at the time of publication. These changes and updates should enhance your use of this book and enjoyment in rockhounding in Idaho.

Idaho is a vast treasure chest of gems, minerals, rocks, and ores. Mineral collectors and rockhounds probably spend more time outdoors in Idaho than any other group except, perhaps, fishermen. The lure of the agate, quartz crystal, or topaz gem, along with the pure enjoyment of all that nature has to offer, keeps collectors returning time after time to favorite and new collecting sites. This book is written with the idea of sharing these collecting sites with others who already share the author's love of the outdoors.

Rockhounds, and others interested primarily in agate, petrified wood, and other cutting material, have probably been the greatest number of individuals searching for specimens in every region of Idaho—its mountains and its deserts. In recent years, there has been an increased interest in the mineral specimens and crystal treasures of the state too.

There are many localities, each with its own type of specimens, for every interest. In this book, you will be directed to forty of them. Each one has its own collecting conditions, and unique access. All of these are described in the text and maps so that you will be prepared when you get to the localities.

Collecting tools required vary from one locality to another, for there are several types of occurrences. When collecting agate, jasper, and petrified wood in the desert areas, it is often necessary only to walk along and pick up the pieces on the surface of the ground. At other locations, the agate, crystals, or other material is mixed with the soil or stream gravels. For these, a shovel and pick, along with a screen, will be most useful. When collecting agate or other cutting material that occurs as seams or nodules in solid rock, or various crystals that occur in cavities in solid rock, it will be necessary to use heavy hammers (two to four pounds), or perhaps an eight-pound sledge, various points, and chisels. Be sure and take good care of your equipment. Good handles are necessary to decrease the chance that you will travel all the way out to the locality only to break the tool. Also, keep those chisels and points sharp.

For safety purposes, always wear goggles when breaking rock and using hammers and chisels. Long sleeve shirts, pants, sturdy boots, and gloves are also necessary for protection. The rocks are always hard, and often break with sharp edges. Most collecting involves a certain amount of hiking and walking through brush, which can take a toll on bare arms and legs. Protect yourself and make each collecting trip more fun.

Unfortunately, the outdoors also harbors other creatures. Mosquitoes are common in the mountains and other areas where there are streams and lakes. Ticks are common in the deserts and forested areas alike, especially in the spring and early summer. Don't forget your insect repellent to keep most of these pests away.

Property ownership changes are common, so be aware. All of the localities were believed to be open to collectors at the time this book was prepared. Any lands can be sold or traded, or on public lands, claims may be staked, so the ownership of the land or the minerals can change. This book does not give the reader permission to trespass on private lands or mining claims for the purpose of collecting specimens. Many land owners welcome collectors who are courteous, and most owners of mining claims do not discourage amateur collecting. To protect your own interests as well as those of the land or claim owner, permission should be asked first from all private land owners and mining claimants.

Hobby collecting is permitted on the public lands—those administered by the Bureau of Land Management and by the U.S. Forest Service—but not all public lands are open to collecting. The status of these lands change, so you may want to check with the Bureau of Land Management or Forest Service to inquire about the status of the land. One should be a good citizen on these lands and not make any unnecessary or large excavations. It may be a good idea to check with the administrative office of the land management agency before doing any work beyond a "casual" amount of rock breaking or surface disturbance.

# TOOLS

The tools you carry are the most important equipment that you will have for collecting. Certainly, you may also need some camping equipment, food, clothes, and other items; but to actually do the collecting, you will need the proper tools. Experience will teach you what tools are the best for you for each type of collecting. The list below will help you select your beginning equipment.

## HAMMER AND SLEDGES

The hammer is the most useful tool for much of the collecting. The geologist's pick is popular with collectors and has many uses. The pick has a point on one side and a square head on the other. The point is useful for light digging, loosening a rock half-buried in the dirt, and for picking apart soft rock. The square head is good for chipping pieces off of rocks in order to expose a fresh surface to see what is inside.

For breaking larger rocks and driving large chisels, choose two- to four-pound hammers. These are most useful for much of the chisel work you may do and for breaking rocks that aren't too large. If the rocks are large, choose eight- to twenty-pound sledgehammers. They can drive larger chisels into cracks and readily make small rocks out of big ones. Don't drive small diameter chisels with heavy hammers, you will bend the chisel before accompanying any rock breaking.

## CHISELS AND POINTS

There are several sizes and styles of chisels that work well for collecting specimens. Cold chisels of various sizes are useful for widening cracks. Use small chisels on small rocks, and large ones on large rocks with a large hammer or sledge. For removing small specimens, especially small clusters of crystals, use a narrow, thin chisel, and a lightweight hammer. For making cracks in rocks, you may want to use points, not flat chisels.

## PRY BARS

Bars come in all sizes like the other tools. Smaller ones are useful on small rocks, and larger ones are more useful for larger rocks. All sizes of large bars are available, but for the small job, there are two bars that are most useful. One of these is the Gad-Pry by Estwing (they also make all sizes of hammers). The Gad-Pry has a point on one end and a 90° bend with a short chisel on the other. Another useful bar is an old nail bar, some collectors cut the curved end off so they can hammer on it.

## DIGGING TOOLS

When collecting agates, petrified wood, and other specimens in the soil and stream gravels, the most useful tools are shovels and screens. You may need a long-handled shovel when a lot of digging is to be done, or deep holes are necessary. For a small amount of work, or for carefully searching the dirt, use a small folding shovel. The most useful screen is generally one with a one-fourth-inch mesh in a wooden frame. You may also need a pick to loosen compact soil.

## CARE AND PROTECTION

When using tools and equipment, especially when breaking rock, always wear goggles or use other eye protection. Leather gloves may be necessary to protect your hands, and a good pair of boots are often important to protect your feet. Be careful and protective of others too. These are only a few suggestions to avoid injury. You are responsible for having a safe collecting trip.

# USEFUL ADDRESSES

The following two are sources for geological publications, including maps and topographic maps.

**STATE GEOLOGICAL SURVEY**
Idaho Geological Survey
Morrill Hall, Room 332
University of Idaho
Moscow, ID 83843-33014
(208) 886-7991
www.uidaho.edu/igs/igs.html

**U.S. GEOLOGICAL SURVEY**
Earth Science Information Center
Room 135, U.S. Post Office
West 904 Riverside Avenue
Spokane, WA 99201
(509) 368-3130 / Fax (509) 368-3194

## U.S. FOREST SERVICE

There are many Forest Service offices in the state. Each forest has a supervisor's office, and there are several ranger district offices in each forest. The supervisor's addresses are given below. Any Forest Service office can give you the address of any other Forest Service Office if you need a ranger district office. The Forest Service can offer information about access, land ownership, and provide maps and camping information.

National Forest Offices are located in several Idaho cities:

**SOUTHERN IDAHO:**

Boise National Forest
1249 S. Vinnell
Boise, ID 83709
(208) 373-4100

Caribou National Forest
Federal Building, Suite 282
250 South 4th Avenue
Pocatello, ID 83201
(208) 236-7500

Challis National Forest
Highway 93 North
HC 63, Box 1669
Challis, ID 83226
(208) 879-2285

Salmon National Forest
Forest Service Building
Highway 93 South
Box 600
Salmon, ID 83467
(208) 756-5100

Targhee National Forest
420 North Bridge Street
P.O. Box 208
St. Anthony, ID 83445
(208) 624-3151

Payette National Forest
800 West Lakeside
P.O. Box 1026
McCall, ID 83638
(208) 634-0700

Sawtooth National Forest
2647 Kimberly Road East
Twin Falls, ID 83301-7976
(208) 737-3200

**NORTHERN IDAHO:**

Bitterroot National Forest
1801 North 1st Street
Hamilton, MT 59840
(406) 363-3131

Idaho Panhandle National Forests
3815 Schreiber Way
Coeur d'Alene, ID 83815
(208) 765-7223

Clearwater National Forest
12730 Highway 12
Orofino, ID 83544
(208) 476-4541

Nez Perce National Forest
Route 2, Box 475
Grangeville, ID 83530
(208) 983-1950

## BUREAU OF LAND MANAGEMENT

Similarly, there are many Bureau of Land Management offices to administer the Bureau of Land Management lands;

Bureau of Land Management Offices are listed below:

### SOUTHERN IDAHO:

BLM - Lower Snake River District
3948 Development Avenue
Boise, Idaho 83705
(208) 384-3300-Information

BLM - Upper Snake River District
1405 Hollipark Drive
Idaho Falls, ID 83401-2100
(208) 524-7500

BLM - Malad Field Station
138 South Main
Malad City, ID 83252-1346
(208) 766-4766-Information

BLM - Shoshone Field Office
400 West F Street
P.O. Box 2-B
Shoshone, ID 83352-1522
(208) 886-2206-Information

BLM - Jarbidge Field Office
2620 Kimberly Road
Twin Falls, ID 83301-7975
(208) 736-2355

BLM - Burley Field Office
15 East 200 South
Burley, ID 83318
(208)677-6641-Information

BLM - Pocatello Field Office
1111 North 8th Avenue
Pocatello, ID 83201-5789
(208) 478-6340

### NORTHERN IDAHO:

BLM - Upper Columbia - Salmon/Clearwater District
1808 North Third Street
Coeur d'Alene, ID 83814-3407
(208) 769-5000

BLM - Salmon and Challis Field Offices
Highway 93, South Route 2
Box 610
Salmon, ID 83467
(208) 756-5400

BLM - Cottonwood Field
House 1, Butte Drive
Route 3 Box 181
Cottonwood, ID 83522-9498
(208) 962-3245

# Overview of Locality Sites Map

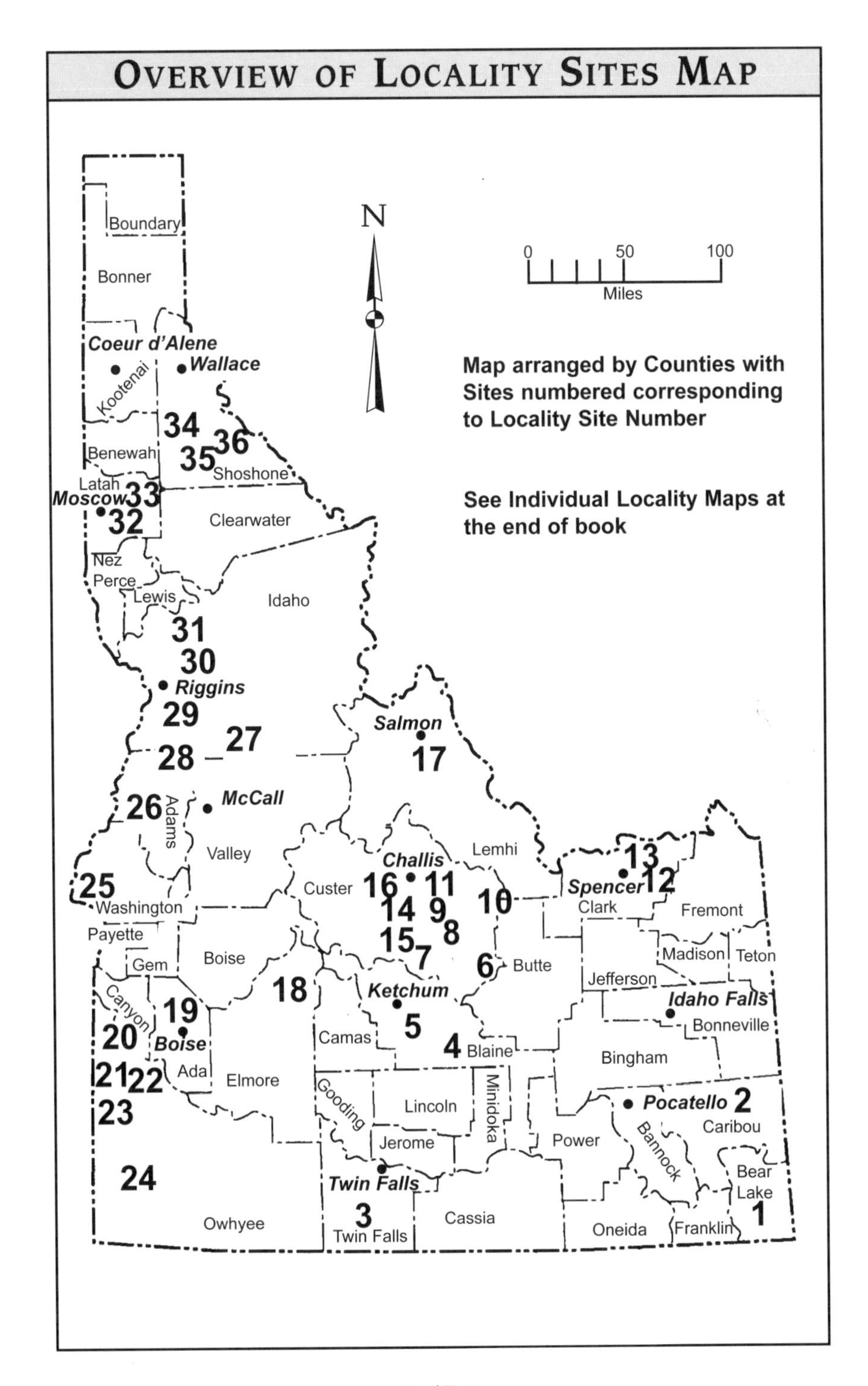

## 1. PARIS CANYON - MALACHITE IN QUARTZ

The lapidarist looking for an unusual material to cut need go no further than this remote site in the southeastern corner of Idaho. The malachite forms thin seams and layers on white and colorless quartz, often coating white quartz breccia fragments. In many pieces, the quartz is a reddish-brown jasperoid. Some very nice pieces with interesting patterns and contrasting colors can be found. Red iron oxides are common. Some red, green, and white specimens are available, especially pieces with gray, brown, red, and white quartz intermixed with the green malachite.

The deposit is an old overgrown mine and dump. The road passes over the top of the old dump with the small open pit nearly hidden in the woods on the left, and the dump extending down the hill to the right. Young trees and brush have grown up over all the old workings.

Pieces of the cutting material can be found in and alongside the road, in the pit, and the dump. There are a couple more small pits and dumps extending down the hillside for a hundred feet. Material can be collected by picking it up, breaking it off large pieces (to more than a foot across), and by digging it out of the dump.

This is National Forest land, but there is private land farther down the road and adjoining the access road. Watch for signs to avoid the private land.

## 2. CHINA CAP – FLUORESCENT GLASSY RHYOLITE

This locality will probably only be of interest to those who collect fluorescent minerals—those minerals that glow a different color under ultraviolet light than under normal white light. The material is a vitric (glassy) rhyolite that makes up a large cinder cone a couple of miles south of Blackfoot Reservoir, about twelve miles north of Soda Springs. China Cap is the middle cone of three cones in a northeast-southwest trending line on BLM land.

In the southwestern part of the cone, a pit has been excavated in the rhyolite. The fluorescent rhyolite is common and makes up several layers in each piece of the rock, but not all of the rock fluoresces. The color of the rhyolite is gray and the fluorescent color is an orangish-cream color. Also, for those who collect micro-minerals, some cavities in the rhyolite contain tiny crystals of topaz along with tridymite, quartz, and hematite. The topaz crystals are unusually elongated and up to about two millimeters in length.

## 3. RABBIT SPRINGS – AGATE AND THUNDEREGGS

About four and one-half miles north of the Idaho-Nevada border on Highway 93, there is an interesting agate and thunderegg locality at what was the BLM Rabbit Springs Rest Area. However, the restrooms have been removed. Agate as seams and partially filled geodes occur in an extensive quarry a half-mile southeast of the rest area. Only a little of the agate is of good quality, but some of the thundereggs have a good form.

**Agate thundereggs at the Rabbit Springs Roadcut on Highway 93**

Immediately north of the access road to the rest area, the highway goes through a cut in red and black volcanic rock. Primarily on the west side, this rock in the top portion of the outcrop is loaded with small (one to three inches) thundereggs. These are not the typical thunderegg filled with agate. These are hollow and are lined with a bluish and white agate material, but they have the classic star-shaped cross section. They are wonderful examples of the thunderegg, showing the form well, and the "bubbly" agate filling is interesting. Of great interest to many, this agate fluoresces a light reddish color. The thundereggs are exposed on the surface of the ground west of the highway, and can be easily collected. The Highway Department has reconstructed the cut so that it is rounded. The thundereggs litter the ground and can easily be picked up.

## 4. MULDOON – PLUME AGATE AND BANDED AGATE

This area produces excellent, green moss agate (some of it has a banded central portion), plume agate, and banded agate. Large pieces up to several inches can be found. This material, and other agate, occurs south of Muldoon over a large area. The area I checked out is the first ridge east of the road, across the creek from the old town site of Muldoon. Anywhere on the ridge for about one and a half miles is good collecting. There is a bridge at the sheep corral which can be used to cross Muldoon Creek to collect material on the ridges on the west side of the creek.

Agate fragments can be found on these hillsides scattered amongst the grass and sage brush. The pieces and nodule sections are mostly about one inch across, but can be found larger than five inches. Most of it is agate, generally banded, typically with the colorless, white, and light blue coloration, but it can also be pink. Moss or plume agate pieces are uncommon, but some excellent pieces have been found. You have to look hard for these, but they are worth the extra effort. There are pieces of volcanic rock with seams of blue and white agate, and occasionally, an agate nodule with quartz crystals in the center can be found. It is only necessary to walk along and pick up pieces off the surface.

There is private land in the area, so watch for signs. The hillside and much of the area is BLM land and is open to public access.

## 5. COLD SPRING CREEK – AGATE AND JASPER

This is a good locality for easy access and easy collecting. Pieces of agate and colorful jasper can be found on the hillsides and valley floor along the road, and in the creek for several miles. In the creek, from where the road crosses the creek upstream, it is possible to find pieces of agate and colorful jasper (browns, red, greens, black) up to fist size. In the flats along the road and creek, there are lots of pieces of agate and jasper. The agate varies from white to blue, to yellowish, gray and smoky, banded, small nodules, and irregular pieces.

Material has been reported here and there along the road from this site all the way to Muldoon Summit towards Bellevue. There is a large area to look over and keep the collector occupied for some time. Access is easy, and this is open BLM land.

Back down the Little Wood River, there is a campground near the dam. Across the river on the hillsides to the immediate west of the dam, there are small pieces of yellowish agate and some tube agate. Small, yellowish tubes, sometimes tinged with red, occur in colorless, white or yellowish agate.

## 6A. MACKAY MINING DISTRICT – GARNET, CALCITE, AND CHRYSOCOLLA

The mines that are up in the mountains west of Mackay have many dumps that can be fun to pick around in. There are also many old mine buildings that are very photogenic. The sightseer and history buff (especially the lover of ghost town-type ruins) will like this area. On the way up to the mines,

**Mill of the Empire Mine in the Mackay Mining District**

the towers for the tram, which was used to carry ore down the mountainside to the smelter, can be seen. A good road provides access to the dumps and the huge old wooden ore bins of the Empire Mine, and just beyond it to another mine. Look down the mountainside to see a rare sight of an old wooden ore car trestle projecting out of the trees to nowhere, with a dump at the end. By driving other roads in the area, it is possible to find more mining history.

When in this area be careful. Like most mining areas it has open pits and shafts that should not be entered and are especially hazardous to children. The lands are National Forest and BLM lands with some private land. Most of the public lands have mining claims located on them so some may be off-limits to collectors. Watch for signs in this area of mixed land ownership.

The mine dumps can provide many specimens of interest. The dumps at the Empire, and other mines in the immediate area, can provide specimens of small orangish-brown grossular garnet, black magnetite crystals, white calcite crystals, gray molybdenite, and large rock surfaces coated with beautiful layers of blue chrysocolla and green malachite. Collecting can be done by searching the mine dumps. A hammer and a chisel may be useful.

## 6B. MACKAY – JASPER

This small prospect along the road to Alder Creek provides a jasper locality that is different than most. The jasper forms large masses outcropping in a small prospect pit beside the road, and as small to large pieces scattered along the road and hillside around the prospect and a quarter-mile farther up the road.

The jasper is quite colorful and, when cut, shows patterns of swirls, layers, patches, etc. The colors include black, browns, reds and nearly white. There are many holes in some pieces, but large solid areas are common. This material is quite colorful and takes an excellent polish.

## 7. LITTLE FALL CREEK – QUARTZ AND MOLYBDENITE

Near Trail Creek Pass, between Mackay and Ketchum, there is a scenic narrow canyon that produces nice specimens of drusy quartz. Most of the material is coarse with crystals up to about a half-inch long. It can be found in plates to about six inches across. The main locality is a talus slope across Little Fall Creek a short distance after the road crosses the creek the first time. This talus is obvious, as the road is constructed through it.

The quartz specimens can be found lying in the talus and can easily be picked up, but for some, you may have to chisel them free from the rock. Similar material can be found in other places in the canyon, all the way up to where the canyon forks.

Molybdenite occurs as small hexagonal plates in a prospect along the road past the large talus slope. The small benched area is on a small side road. It is necessary to break rock to expose the small cavities with molybdenite crystals up to about one-fourth inch across.

**View down Little Fall Creek Spur to the Molybdenite area**

The road is narrow and rough, and can be traveled with difficulty with a car up to the first creek crossing. It is only a short walk from there, and the talus can be seen across the creek. A pickup or four-wheel drive vehicle is necessary to travel the rest of the way. The road continues a distance up the canyon through a meadow area. The snow doesn't melt out of this area until late June or later some years. After that, water may still be high in the creek crossings. This is National Forest land, but is probably best for late summer collecting.

## 8. CHILLY CEMETERY – CALCITE

This is another easy access locality from which one can collect some nice calcite that can be cleaved into perfect rhombs. The calcite is milky to rarely colorless, and some have gray inclusions. Individual pieces can be found up to three inches across or better. The locality is visible from the Trail Creek Road as two bright red small prospect pits and dumps on a small hill above the historic Chilly Cemetery.

You will need to dig in the red dirt of the dump to find the pieces of calcite. These can then be cleaved with a hammer to make very sharp rhombs of good size. As at other mining areas, stay out of the workings (even though shallow), and use caution around all steep banks and cuts.

These are BLM administered lands, but mining claims may be located on these prospects at times.

## 9. LONE PINE PEAK – AGATE

This agate deposit southeast of Challis is another easy-access collecting site that has not been picked over. It produces some nice banded agate in several colors—blue, white, and pink. Individual pieces, nice-sized for cutting, are up to four inches. Jasper is reddish-brown to red, sometimes with streaks of green. I have also seen small pieces of solid green jasper.

This is a dry area, with only a sagebrush cover. Pieces of volcanic rocks and agates are exposed on the surface, so it is only necessary to walk along the valley and hillsides, picking up what you want. Good collecting can be found on the middle knob of the three major high points of the ridge north of the road. There are large pieces in the rock outcrops, in the talus below these, and on the hillside. Good material can be found along much of this ridge. I found some of the best material near the truck parked alongside the road, just before the saddle.

In the spring, antelope, sage grouse, and jack rabbits can be seen in the area. The road is not maintained, but is a good truck route that can be managed with care by most people in cars. There are a couple of mud holes that could be a problem after a rain storm. You should be familiar with driving unmaintained roads to access this site which is on open Bureau of Land Management land.

## 10. LIME CREEK-PASS CREEK – GEODES, AGATE, AND QUARTZ

This Lime Creek is a tributary of Pass Creek which provides access from the Mackay-to-Arco Highway over to the Pahsimeroi Valley. The first part of Pass Creek goes through a narrow limestone canyon, which is very scenic. It is also geologically interesting; one can see the bedding of the limestone which is often tightly folded.

Lime Creek is a small tributary on the west side that has a large area with agate and jasper of many colors and types. It occurs in seams and nodules and as nodules with quartz crystal centers. In about an hour, I was able to pick up several pieces of agate and jasper. The agate is light blue or colorless in the most part, but is also banded with blue, white, and blue-gray; some is light pink. Other nodules have a bluish-gray rind with quartz crystals in the center. Pieces of jasper are usually quite colorful with swirling patterns of light brown to dark brown, yellow, black, red, and some greenish color.

Most material is found by walking the steep hillsides and picking up the pieces. It is also possible to find the material in the volcanic rock outcrops and to collect it with a hammer and chisel.

## 11A. LIME CREEK – AGATE

This locality is only a short distance southeast of Challis, and is easily accessible (don't mistake it for the other Lime Creek much farther to the southeast!). Agate can be found on the hillsides, west of the road and creek, as pieces from about an inch up to three inches or more across. The color varies from colorless, to white, to blue-gray, and blue. They are of a solid color or banded. Some have complete cross sections of nodules showing the circular banding pattern. There are also pieces of reddish-brown jasper.

Access to this part and the next part of Lime Creek is open on these BLM lands. The road generally can be passable by car, but may not always be in good shape.

## 11B. LIME CREEK – STILBITE AND AGATE

At one and nine-tenths miles up the same road as the previous locality, a narrow canyon runs off to the west. It is only

**Stilbite in agate-lined cavity from Lime Creek**

a short hike (about one hundred and fifty yards) up this canyon into an area of volcanic breccia which contains numerous seams of blue to greenish blue agate. Some of these are translucent, and can be cut into nice stones. There are numerous cavities lined with drusy quartz and occasionally containing very nice crystals of white stilbite. Most of the stilbite is less than an inch long, but some are larger. There is also a little heulandite and some calcite. Uncommon, are larger cavities with quartz fingers with stilbite crystals on them; these are fabulous specimens! Some of the quartz is a second-stage deposit and is a little larger than the tiny drusy lining in most cavities. Sometimes, this quartz forms radiating groups, and occasionally, these crystals may be sceptered. You will need your heavy tools to collect this material.

You will have to climb around the steep canyon sides looking for exposed cavities or for float material indicating a buried cavity. A pick and even a shovel will help to dig out buried cavities.

## 12. CINDER BUTTE – FELDSPAR

Faceters will love these small feldspar crystals. They are mostly a light yellow color, or colorless (light reddish colored pieces have been reported), and can be found up to an inch across. It is possible to find pieces that can be cut into stones of ten millimeters, and pieces two to four millimeters are common.

The feldspar crystals weather out of the lava rocks of the butte and can be found scattered on the ground around the butte and on its sides. It is only necessary to walk along, looking for the sparkle of the crystals. It is also possible to dig in the cinders to find crystals in place. The cinders are loose, so a small pick can loosen and separate them; most collectors will find it satisfactory to simply pick up exposed pieces.

Some of them are excellent, well-formed crystals (good examples of the plagioclase group mineral andesine); most are somewhat rounded and not perfectly formed. They vary from cloudy to completely transparent.

This is BLM land and has been open to collecting. On occasion there is equipment in the pit excavating cinders, so be careful and avoid any mining operations.

## 13. SPENCER – PRECIOUS OPAL

This may be the most popular collecting area in the state of Idaho. Since its discovery several decades ago, thousands of people have come to the fee area east of Spencer to dig nodules of white opal with layers of colorful precious opal.

From Interstate 15 take the Exit 183, go left and a left again at the main street. From there follow the road to the dead end until you reach the Spencer Opal Mine shop. The mine is open on specific weeks, Saturday-Monday, from May 15 through mid-September. Call the Spencer Mine office at (208) 374-5476 for these dates. Collectors must meet at the mine office in Spencer and take a bus furnished by the mine to the diggings (at 7,000 feet elevation in the Bitterroot Mountains) or take their own cars to the mine (the mine road is not recommended for motor homes or low cars). Digging hours at the mine are 8:30 a.m. to 4:00 p.m. The fee is $25 per digger per day, and is payable at the mine. For a brochure, send a self-addressed stamped envelope to Spencer Opal Mines, HCR 62 Box 2062, Spencer, Idaho 83446 or visit www.spenceropalmines.com. From October through April, write to P.O.Box 521, Salome, Arizona 85348.

Collecting is done by breaking the hard volcanic rock with large hammers and chisels to remove the opal nodules, or by looking for nodules and pieces of nodules in the debris from the commercial efforts. Bring water and food, gloves, and both rain gear or sun protection. The opal is high quality, and generally with thin seams, which make nice triplets, and a small quantity are of sufficient thickness to cut solid precious opal stones.

## 14A. ROAD CREEK – AGATE

Road Creek, a tributary of the East Fork of the Salmon River, has been known for its agate and jasper for many years. It continues to produce good material for the collector who is willing to walk the hillsides of the valley and pick up the pieces from the surface.

**Agate exposed in weathering basalt from Road Creek**

This locality on Road Creek produces some very nice banded agate nodules. They are small and most are approximately an inch across, but they are very nice. The colors range from colorless, white, gray, to blue-gray. They can be found in the upper parts of the small grassy-floored canyon. The hillsides around this canyon also have small pieces of agate and the author found a small piece of carnelian and one thin-shelled geode that had a nice pink agate layer in colorless agate.

Don't limit your agate hunting in this area to this spot; there are a lot of ridges and exposures of these same volcanic rocks in this area. I have seen a few pieces of amethyst lined geodes from these canyons.

## 14B. ROAD CREEK – AGATE, NODULES, AND JASPER

While checking out this area, I found an alluvial fan at the mouth of one side canyon that appears to have been untouched by collectors. The surface of the fan is littered with pieces of agate, nodules, and jasper. The agate is in irregular nodules and seams, much of it light blue and blue gray. Some of it has bands of white, gray, blue-gray, and blue. There are nodules that have banded agate rinds with quartz crystal centers. A few jasper pieces were also found.

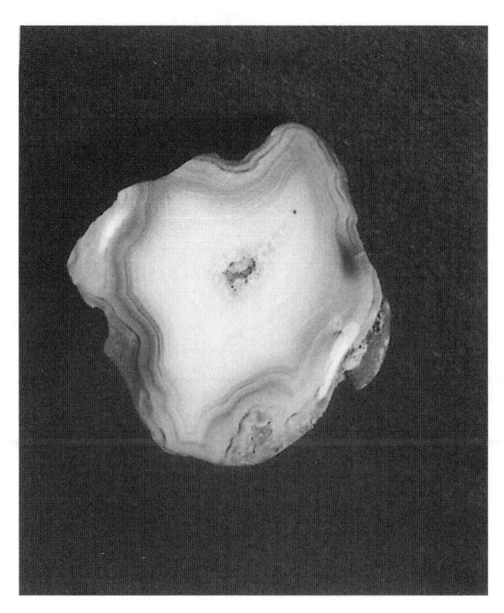

**A polished banded agate from Road Creek**

The agate material can be found in the greenish-brown outcrops that are about half-way up the "V"-shaped mouth of the canyon. This material is abundant in the rock in some areas, and can be easily removed with a hammer and chisel. Not all of the material has come from these rocks, but they do account for the abundance of it on the surface of this alluvial fan. Similar material can be found up the canyon, where the "V" narrows, and on the hillsides above.

This area is easy collecting, especially for small pieces for the tumbler. There is much material here, in pieces up to about four inches. With the great variety, there is something for everybody.

## 15. HERD CREEK – AGATE AND JASPER

When you are in the East Fork area, you can drive a little farther south beyond Road Creek to another good agate and jasper area, this time along Herd Creek. Collecting conditions are similar to those on Road Creek, but the canyon here is mostly not as rugged. There is good collecting in several areas along the road, starting at about a mile up. Materials include banded agate with bluish gray, white and gray colors predominating. There are pink layers, yellowish areas, and some carnelian. Pieces of jasper, typically reddish-brown to red, and some with scenic patterns were seen. A few of the agate nodules had hollow centers lined with quartz crystals.

You don't need your collecting tools here, just walk along the hillsides and look for the agate and jasper pieces on the surface. Herd Creek runs water into mid-summer some years; the gravel bars of the creek are good areas to hunt for agates.

The East Fork Road is quite scenic and provides access to the White Clouds Peak section of the Sawtooth National Recreation Area. You will find trails to hike and places to camp. This is a great area in which to take a scenic drive.

## 16. GARDEN CREEK – FLUORITE

A small area west of Challis was prospected for fluorite with a small amount of production from a couple pits and prospects. At the larger pit shown on the map, fluorite primarily occurs as a crystalline crust of colorless to light lavender to pinkish color. The crust lines the west wall of the pit and can be removed in large slabs about one and one-half inches thick with a cubic crystalline top. There are cavities in some of the rock in the pit that has larger crystals.

To the south of this pit, up the ridge, and over the south side, there are small prospect trenches. Better fluorite crystals can be found in these. It's necessary to break hard rock to get them; try the boulders that are lying around. The crystals are sometimes brilliant, and some of them have a good medium lavender color.

## 17. WILLIAMS CREEK – JASPER

This is just about the easiest collecting possible! Drive up the road, park, get out, pick up pieces of jasper, get back in car, go camping. Jasper forms veins up to about an inch thick in the light colored volcanic rock of the north canyon wall. The rock forms a large talus slope reaching from the road up to the rock outcrops a few hundred feet above. There are lots of jasper in some places, especially the last area of the talus up the road, ending where the dark rock begins.

The jasper is dominantly a nice rich brown, slightly reddish with minor amounts of light brown, buff and creamy white. Some of it forms a pattern of the light colors as bull's-eyes in the darker colors. This material is very hard and polishes to a high gloss. The largest piece I saw (and left behind) was a slab about ten inches long, four inches wide and an inch thick. The BLM has posted "do not remove mineral materials" signs, but these are to stop the removal of large quantities of the talus rock, not the personal collecting of small quantities of the jasper.

## 18. DISMAL SWAMP – SMOKY QUARTZ AND TOPAZ

This is an old, well-known locality, that has even seen a little commercial mining activity. What is all the attraction? How about smoky quartz crystals, often faceting quality up to more than six inches in length, or nice topaz crystals, sometimes facetable, to about two inches in length? Most of the topaz is colorless, but some crystals have a light pink color.

Access to this locality is easy, but it is a long enjoyable drive. The locality is marked as a "Rockhound Area" on the Boise National Forest map. National Forest Maps can be obtained by contacting the main office at (208) 373-4100 (see "Useful Addresses" section). This area is accessible from several directions, including the Middle Fork and South Fork of the Boise River, or from the Rocky Bar-Featherville Road. This is a spectacular, scenic area, whether you are either driving along the forks of the Boise River in narrow canyons, or along ridge tops at 6,500 to 7,500 feet elevation. Naturally, it is not accessible from about October to late June, due to snow cover.

Collecting requires digging in a stream or swampy area. The best tools are a shovel, pick, and a screen. Naturally it is most comfortable during the summer months, when the weather is warm and sunny.

Camping can be done near the collecting area, along the main road above the locality, or along the Middle Fork of the Boise River. There are many roads to drive in this area, and some spectacular country to see.

## 19. LUCKY PEAK DAM – ZEOLITES AND CALCITE

If the family is looking for a good spot to go for a picnic, some swimming, boating, and all those other activities done around a lake, but also do some mineral collecting, then this site would be hard to beat. This locality is at a picnic area at the Lucky Peak Dam Reservoir. The best material is below the road, under the picnic tables (and is underwater in the spring). There also is some chabazite and calcite in the roadcuts.

In the small cavities in the reddish to gray basalt, the zeolites line or fill cavities. Chabazite is common as tiny colorless crystals. Phillipsite forms tiny crystals lining cavities or as balls to nearly one-fourth inch across on thomsonite. Thomsonite forms colorless to gray (rarely bluish) crusts and tiny balls. Levyne is scarce, but can be found as plates to nearly one-fourth inch across. Calcite forms scalenohedral

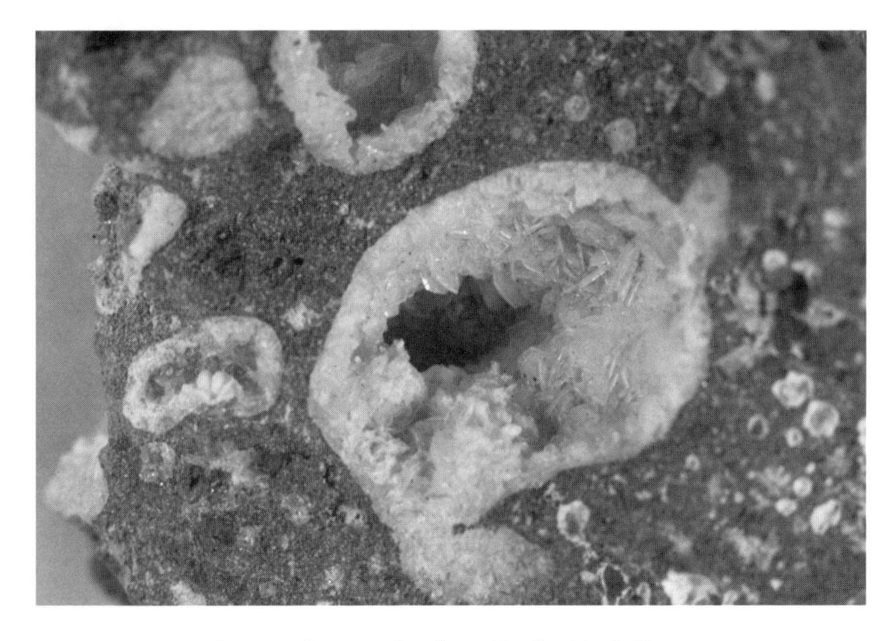

**Levyne in a cavity from Lucky Peak Dam**

crystals up to around an inch long, and is most common in the roadcuts.

The rock here is fairly easy to break, but it is a picnic/recreation area, so be considerate of others and don't break large amounts of rock in the picnic site.

## 20. SOMMERCAMP ROAD – JASPER

This is another easy access locality on BLM land that produces very nice cabbing material. This jasper is scattered over the surface of the ground and requires no digging. The best access is where the fence along the highway makes a jog; there is parking there too. The pieces of jasper are mostly one to three inches across and are of many colors: lavender, light blue, yellows, white, browns, nearly red, and greens.

All that is necessary to collect this material is to walk around the area shown on the map and pick up the pieces. Some of them have a white coating and can be easily overlooked if there is no exposed corner. For those who like to dig, go to the eastern edge of the flat area where it drops off into the valley and dig along the edge. Many pieces of jasper can be recovered this way. It is also a good idea to look here for pieces that have been dug out and left behind by others but are now washed off by rain.

The jasper is on the sloping surface of lake sediments. The sedimentary layers can be seen in the roadcuts to the immediate east. The jasper probably originated in the volcanic rocks to the south and have been carried downslope by erosion.

## 21. GRAVEYARD POINT – AGATE

The plume agate from Graveyard Point, is some of the finest in the state. White plumes occur in a colorless agate matrix, and has been long sought after by Idaho gem cutters. The agate occurs in seams, and requires hard work with hammers, chisels, and bars to collect, but it is well worth the effort.

The map shows the access to the original Graveyard Point locality, a little ways beyond a marker erected by local rock clubs. There are other collecting localities in this area, so it is possible to enjoy a several day trip and collect agate and jasper of several colors. The map shows the road access to an area to the south, which is also known as Graveyard Point. Much of this area has been claimed for commercial production of the agate. Permission can usually be obtained for individuals who would like to collect with hand tools. Inquire at the T.E.P.E. Rock Shop, Marsing ID, 83639 at (208) 896-4311 or at Stewart's Gem Shop, 2618 West Idaho Street, Boise, ID 83702 at (208) 342-1151.

## 22. COAL MINE BASIN – PETRIFIED WOOD

The Coal Mine Basin petrified wood area is small but interesting. The wood can be found on the north side of the road, at the saddle and farther east in the steep canyon walls. Most of the wood is dark to black inside with a white outside. Much of the white outside is soft and crumbly. The dark interior is hard, shows the grain, and polishes very well. The wood is replaced by opal, and like much opal, it begins to fracture and crumble when exposed on the surface of the ground for many years.

At the saddle shown on the map, pieces of wood can be found where logs have been exposed by erosion and are weathering. A little pick and shovel work will expose large pieces of wood. To the east, it is necessary to walk along the canyon side to find limb sections on the surface of the ground. There is interesting, colorful opal in large pieces to the south on the hillsides. This is open, barren BLM land.

## 23A. McBRIDE CREEK – PETRIFIED WOOD LEAF, AND FOSSILS

There are two good distinctly different localities along McBride Creek for petrified wood; both are on BLM lands. The first is less than a mile from Highway 95 and is an easy hike from the access road, just north of the Oregon-Idaho border. From the point marked on the map, you can look northerly up a small wide valley and see the light colored layers of the lake sediments and volcanic ash layers. The middle layers are white and are quite obvious.

The petrified wood occurs in the upper gray layer which forms the tops of the ridges. It is mostly altered to clay, so is very difficult to work in when wet, and is also quite hard to dig in when dry. From the road, look up this valley and you will see a white topped knob in the middle. If you hike up to the left of this knob and up onto the ridge, you will find a lot of pieces of petrified wood on the surface. They are also quite abundant along the valley to the west. The ridge of sediments runs for about two miles to the north, so there is a large area in which to look for the wood.

If you look around the north side of the little valley from this first small ridge, you can see pieces of the wood on the ground forming small circular areas. These are the exposures of buried

**Petrified wood found on the surface from the McBride Creek area**

logs. The material is partially opal and falls apart under surface conditions. If you dig at these sites, you may expose large pieces of wood, but it is not known how deeply the fracturing of the material goes. Digging is difficult, and no one has dug out a log or large piece of the wood to the author's knowledge.

The wood is black to dark brown when fresh, although some of it is pure white opal. The wood grain is well preserved in most of it. As it lies on the surface, it weathers to a nearly white color on the outside and becomes softer and loses some of the wood grain. Surface pieces vary from less than an inch to about four inches long, six inches wide, and ten inches thick.

That knob in the middle of the first valley is a site for leaf fossils. The gray layer below the white layers is a siltstone that contains an abundance of leaf fossils and pieces of twigs and stems. It is difficult to work, but it is possible to break out large chunks and split them along the bedding to expose the leaves. There are several species of leaves here.

## 23B. McBRIDE CREEK – PETRIFIED WOOD

Go west on McBride Creek Road for two and one-half miles, this road turns sharply left but you will take a dirt road to the right. There are two stream crossings before you get to the main collecting area, and the road is not maintained and can be muddy when wet.

The best area I found is the stretch from about a half-mile north to about one and eight-tenths miles where the road intersects another road and you can continue to the right back out to Highway 95.

Petrified wood is found as float on the flat, slope, and ridge to the west of the road. Wood is generally not easy to find because there is a dense grass cover over much of the ground. Look for the bare spots which are more common on the sides of the ridge. The wood is found in with the rounded pebbles and cobbles. In many spots there were many small pieces of white, brown, black, and gray wood together and an occasional large piece. The largest I found in an hour was two inches long, three inches wide, and five inches thick. It may be productive to dig through the gravel in areas where you find material on the surface.

## 24. SOUTH MOUNTAIN – HEDENBERGITE AND ILVAITE

This area is a scenic area that is well worth the drive. One of the best collecting areas is near the top of the mountain. The map shows a shaded zone that cuts across the road just north of the lookout tower. This zone contains good hedenbergite and some ilvaite. These are uncommon minerals that occasionally occur in world-class quality at this location. On the ridge top, just east of the road is a small pit with some very good hedenbergite in radiating clusters. There are some tiny orangish garnets on the terminations of some of the hedengergite. The zone runs from this pit westerly down the mountain slope to the mines below. Some specimens of quartz and other minerals can be found too.

**Hedenbergite on the surface beside the road near the top of South Mountain**

You can pick up some specimens on the surface of the ground, but it is best to break rock and search for the better specimens. Do not enter the mines; they are unsafe and posted.

## 25. BEACON HILL – AGATE NODULES

This locality has been popular for many years. It is not easily accessible, as it is a five and one-half mile drive up a jeep trail. Although in good condition, it is best traversed by a pickup or four-wheel drive vehicle. The locality is on BLM public land with the access through private ranch land, but collecting has always been allowed. There are four fences across the access road, so be sure and leave the gates as you find them. Due to the nature of the road, it is best not driven in wet weather. Make this a late spring to early fall destination.

The locality is on a ridge top and overlooks the Snake River, which is a wonderful view. The Idaho Gem Club maintains a

**Plume agate nodule from Beacon Hill**

mining claim on the locality, but permits collectors to visit the site for personal collecting only, with hand tools. The material is agate nodules of all sizes from less than an inch up to about three inches. They have blue, gray, white, and colorless bands, and some are hollow, lined with small quartz crystals. There are pieces of agate, white opal, and colorful jasper over much of the area, still lying on the surface.

It is necessary to dig in the rocky soil and move broken surface rock to find the better nodules. I have been told that there has been a recent discovery of better material in the area, perhaps within a half-mile, but I was unable to locate this due to the coming darkness on the day I visited the location.

## 26. SEVEN DEVILS MINING DISTRICT – EPIDOTE, GARNET, AND CHRYSOCOLLA

This is a scenic area that is pleasant to take a Sunday drive in. Access is from Council to Cuprum via a good gravel road. From Cuprum, the road gets narrow and is steep in parts. Although it is passable with a sedan, many people may be more comfortable driving these roads with a pickup.

The map shows several areas where mines occur. Most of these have dumps with brown crystals of andradite garnet. There is also chrysocolla on many of the dumps, and some fine crystals of epidote on some. Some material can be found lying on the surface, and more can be found by breaking rock. Be careful, this is a mining area with open shafts and other underground workings.

**Garnet-epidote from the Peacock Mine in the Seven Devils District**

The upper road can be taken to its end for a spectacular view of Hells Canyon from the Sheeprock Point. This is well worth the drive and short walk on a nearly level trail.

## 27. RUBY MEADOWS – CORUNDUM AND GARNET

The old gold placer workings along Ruby Creek, southeast of Burgdorf, continue to produce nice specimens of corundum and garnet. Occasionally a quartz crystal can be found, and some of the corundum is cuttable, making a nice sapphire gem. The corundum is mostly broken, water-warn crystals of a gray color. An occasional crystal has a blue color, and sometimes is translucent. Other crystals have a pinkish or bronze color and will cut a star sapphire stone of fair quality.

Material can be found by screening the gravels in the piles left from the gold placer work, or by working the gravels along the creek. With a lot of looking it is also possible to find an occasional piece on the surface. The material is not abundant at this locality, but hard work should pay off. The most useful tools are a shovel, screen, and pick.

**A large cut in zeolite-bearing basalt, south of milepost 178 near Pinehurst**

## 28. PINEHURST-HIGHWAY 95 – ZEOLITES

There are crystals of zeolites in the vesicles of the Columbia River basalt seventeen miles south of Riggins in the roadcuts on Highway 95, just north of Milepost 177 for three-fourths of a mile to the north. The basalt is black and the vesicles are lined or filled with white zeolites and other minerals. Most cavities are small, being less than an inch across, but some are larger.

Most of the zeolites are tiny, but good crystals to an inch can be found. There are several zeolites and associated minerals here including: analcime, apophyllite, calcite, chabazite, cowlesite, gyrolite, heulandite, levyne, mesolite, stilbite, and thomsonite. The rock is fairly soft on the surface, but becomes quite hard within about a foot.

At the end of this stretch, there is a new roadcut where rock was removed to repair the road after flooding. This site has a large parking area. There are boulders lying around with zeolites in cavities and good exposures in the cut.

The tools needed include a large hammer, chisels, and long bar. This highway is very busy with much traffic. There are wide parking areas and most of the outcrops are a safe distance from the highway, but be careful not to roll any rocks onto the highway or plug the drainage ditches.

## 29. RUBY RAPIDS – GARNET

This is a nice little collecting area along the Salmon River, east of Riggins. Access is easy; it is only a short drive up the river road from town. The garnets are mostly small, with many around one-fourth inch, and an occasional crystal up to about three-eighths inch. There are a lot of one-eighth inch crystals. These are excellent, red garnet crystal specimens, and some of them can be cut into cabochons, and some can be faceted.

Garnets are easiest to find in the large rocks along the river, below the road. They are common in pockets between boulders, where it is only necessary to pick them up. They also occur in some of the large pieces of gneiss, where they can be broken out. It is possible to find them in the soil above the road, but due to the steep roadcut, digging here is not safe for the collector or the road; although this area was heavily dug.

This is an excellent place for weekend family trips. There are campground farther up the road, and it's possible to camp in spots along the river. The Salmon River Canyon is very scenic.

## 30. SLATE CREEK – ZEOLITES

Analcime and natrolite occur in small cavities along the road east of Highway 95. The best outcrop can be recognized by its appearance of white zeolite spots on the dark gray basalt. It is necessary to break rock with a heavy hammer and chisel to expose cavities. Most cavities are under about one and one-half inches across and contain natrolite needles up to nearly an inch long and small analcime crystals. An occasional cavity will produce an analcime crystal to about three-eighths inch across.

You will need to work with a hammer and chisel, and the rock is hard. There are some good specimens at this location to reward the collector who works hard.

## 31. Cottonwood – Siderite Spheres and Jasper

There is a large (long and high) roadcut in basalt at the exits off Highway 95 to Cottonwood. The basalt has many small vesicles, and some of them contain hemispheres of siderite. The siderite varies from brown to black for those that have been altered to limonite or goethite, but the fresh hemispheres are a light amber or brownish color, often with a dark brown surface that may display an array of colors. Some vesicles will have just a couple hemispheres, but others will have several or be completely coated with them. The hemispheres are up to one-fourth inch or more in diameter. A few cavities contain hyalite opal coating the siderite.

This cliff is nearly vertical and is set back from the highway with plenty of room for parking. You can't miss it. There is nothing else like it in the area and it sits prominently on the south side of the ridge top just south of Mile Post 255. Restrict your collecting to the boulders that have fallen off the cliff. You can break these up with your heavy hammer, a four-pounder will generally do the job.

At the south end of the roadcut, there is a road that goes east. The end of the cut, several yards up this road, contains brown opal and jasper. It can be found as pieces weathering out and lying in the ditch or worked out of the vein in the basalt. Some pieces are green and others have a nice pattern of brown and green colors.

## 32. Mica Mountain – Mica

This is a well-known locality that was mined up until the late 1950s for sheet mica and also produced some beryl. Access is fairly easy, but it is difficult to see the mine. The trees between the mine and the road have grown so tall, that it is possible to miss it. It is best to look for the large open area on the right side of the road as the indication that you have arrived.

Muscovite can be found as pieces scattered over the mine and dump area or can be removed from the mine walls. Good sheets and small books to a couple inches across can easily be found, and larger pieces can occasionally be found.

There is still some black tourmaline to be found here, but it is scarce. Beryl is rare on the dumps now, and not often found in the rock. Crystals are typically a little under an inch across and an inch or more long, but were found up to a foot long and nearly that wide when being mined.

## 33. EMERALD CREEK – STAR GARNET

This area is one of the best known gem localities in the state, producing the state gemstone, and is well-known internationally. It produces some of the finest star garnets in the world. Large crystals up to about an inch across are fairly common, and they have been found up to more than three inches across. Fine quality cabochons, showing four- or six-ray stars can be cut from this material. The effect of the light colored star on a dark purple background is quite pleasing.

The locality is "acquired" land of the St. Joe National Forest, and collecting can be allowed by a permit system only. The Forest Service maintains this collecting locality, open from Memorial Day through Labor Day weekend (Tuesday following Labor Day) each year. Collecting hours are 9:00 a.m. to 5:00 p.m., Friday through Tuesday.

The permit can be obtained from the Forest Service at the locality. It costs $10.00 per adult and $5.00 per child (fourteen and under) per day to dig, this allows five pounds of garnets. If you are fortunate enough to dig more, the excess can be purchased for an additional $10.00 for five pounds or fraction thereof. This is a limited resource, so the regulations restrict you to six days of digging, or thirty pounds of garnets, per year.

To collect these garnets, you will need a shovel, pick, screen, and a container to put the garnets in. Be prepared to dig in a creek, and get wet and muddy. It is a lot of fun for the entire family, and can be quite rewarding. There are a lot of tiny garnets and garnet sands, so if you want some colorful sand or tiny garnets, this is the place to get them.

There is a Forest Service campground four miles downstream from the digging area. Information and maps can be obtained from:

St. Joe Ranger District
P.O. Box 407
St. Maries, Idaho 83861
(208) 245-2531

## 34. FREEZEOUT MOUNTAIN – KYANITE

Kyanite, along with the related mineral andalusite, and with staurolite, is common in many areas of southern Shoshone County and northern Clearwater County. The kyanite occurs in excellent crystals that vary from colorless to light blue to dark blue. They often are partially translucent, and occasionally a small stone can be faceted from a transparent section. Interesting cabochons can be cut from larger pieces. They make excellent specimens for the crystal collector. Large pieces of the rock with the blue kyanite contrasting with the light to dark mica make attractive specimens.

This material can be found as float along the road on the ridge top north of Freezeout Mountain. It can also be found in areas to the east, especially along the road from Moses Butte to Goat Mountain to Black Dome Peak. A small knob on the west side of the Freezeout road contains garnet crystals. The rock is a hard schist, but some large garnets have been found by breaking the rock of this knob with large hammers and chisels.

This material, as well as garnets, can be found in many parts of this region. High elevations and a high relief make up the geography of this area. Be prepared for mountain driving on reasonably good dirt roads. Snow covers the ground much of the year, so plan this as a summer trip.

# 35. O'DONNELL CREEK – QUARTZ

Very good quartz crystals are uncommon in Idaho. There are a few localities that produce some smoky quartz, but they either are closed to collecting for political reasons or produce very little. Amethyst is scarce and is reported to occur in geodes in the agate areas. Clear quartz, or rock crystal, is actually rather scarce as crystals larger than a centimeter or less. One locality that has been known for a long time is on O'Donnell Creek in the Clearwater River area of northern Idaho.

This locality has been open to collectors most of the time, and, on occasion, it has been leased for commercial digging (it is state land). At the time of publication, it has a lease on it, but is open for fee collecting. Contact Gene at (208) 245-9215. The phone is not at his home, so you will have to leave a message. He lives in a trailer a couple miles south of Clarkia (a half-mile south of the Fossil Bowl racetrack, and has tables with crystals for sale along the road. He is setting up to charge a small fee for a day's collecting. To make arrangements, call him and leave a message, or stop at his home. If he is not at home, he specifically says to go on up to the dig site, he will be there. He also has recently obtained the claims at the Hershey Bar roadcut (near O'Donnell Creek) and will be setting it up for fee digging for quartz crystals and garnets.

It's a long drive into this locality. The better road comes in from the south, but this is a crooked major logging truck haulage road. Meeting those big loaded rigs on those corners can be a little exciting! If you choose to use this road, perhaps it would be better to use it on the weekend. The road from the north varies from a good to rough mountain road, much of it not suited for cars. Both roads take off from Clarkia, the southern route branching from the other at the creek crossing at Gold Center. Use the St. Joe National Forest Map to help with this access. Contact the St. Joe Ranger District at (208) 245-2531 to obtain maps. Whichever route used, be sure and take a look at the giant cedar trees at Landboard State Park.

Quartz crystals occur in a zone that extends from above to below the road just before it crosses O'Donnell Creek. Crystals are mostly found as singles and can be found by digging in the dirt. Many people screen the debris to recover the small crystals. Nearly all crystals are brilliant, clear, and sharp. Most are under about one and one-half inches but crystals three inches and larger crystals have been reported. They are not abundant, but hard work will usually reward the persistent collector. Some double-terminated crystals have been found, and a few clusters have been found too.

## 36. Bathtub Mountain – Staurolite

Bathtub Mountain can produce some very fine specimens of staurolite. The best locality is along Trail 100 off the southeast side of the mountain.

The staurolite occurs in crystals to about three inches long in layers in the schist. Crystals are sharp and have a brown color. Most are singles, but 60° and occasional 90° twins occur.

It is necessary to use those hard-rock collecting tools here, including your chisels and four- to eight-pound hammers. The rock can be split along the layering, exposing the crystals, but some of the crystals grow across the layering. Good specimens can be prepared by removing the mica flakes of the enclosing schist. This is a slow and difficult job, but it is possible to expose nice crystals in very good matrix specimens.

The trail takes off from the road at a sharp corner. Follow the trail down a couple hundred feet where you will see outcrops of the schist to your left on the ridge just above the trail. The best staurolite is in a low outcrop that only sticks out above the ground a couple feet. Staurolite occurs in other outcrops too in this area, from the ridge top down to the old road below the trail.

This is a vast area back in the forested mountains. There are several campgrounds in this area, one of which is indicated on the map. Back down on the St. Joe River, there are other campgrounds, and of course this is a great scenic area, whether driving in the mountains or in the canyon along the river.

# 1. Paris Canyon

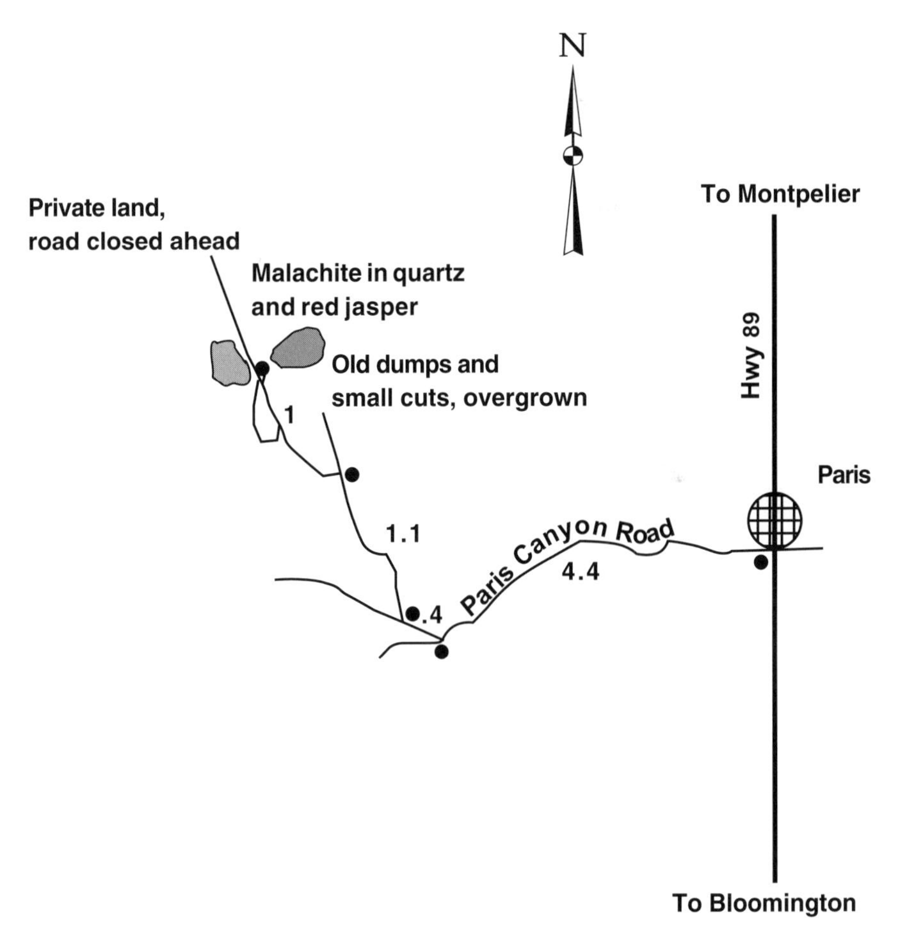

# 2. CHINA CAP

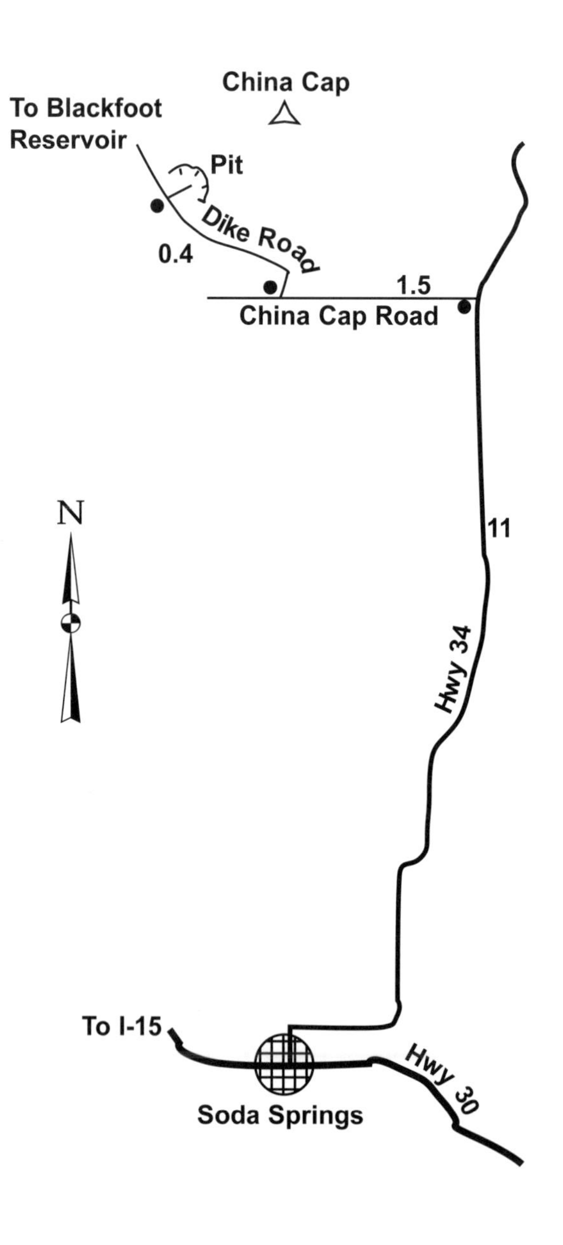
China Cap
To Blackfoot
Reservoir
Pit
Dike Road
0.4
1.5
China Cap Road
N
11
Hwy 34
To I-15
Hwy 30
Soda Springs

# 3. Rabbit Springs

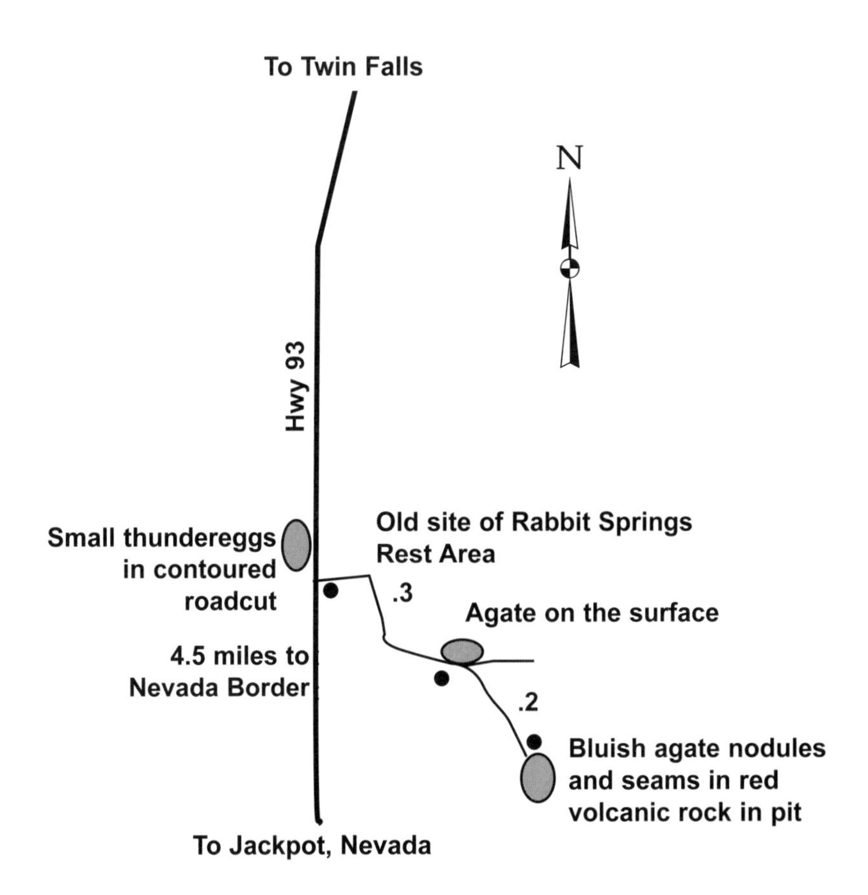

# 4. Muldoon

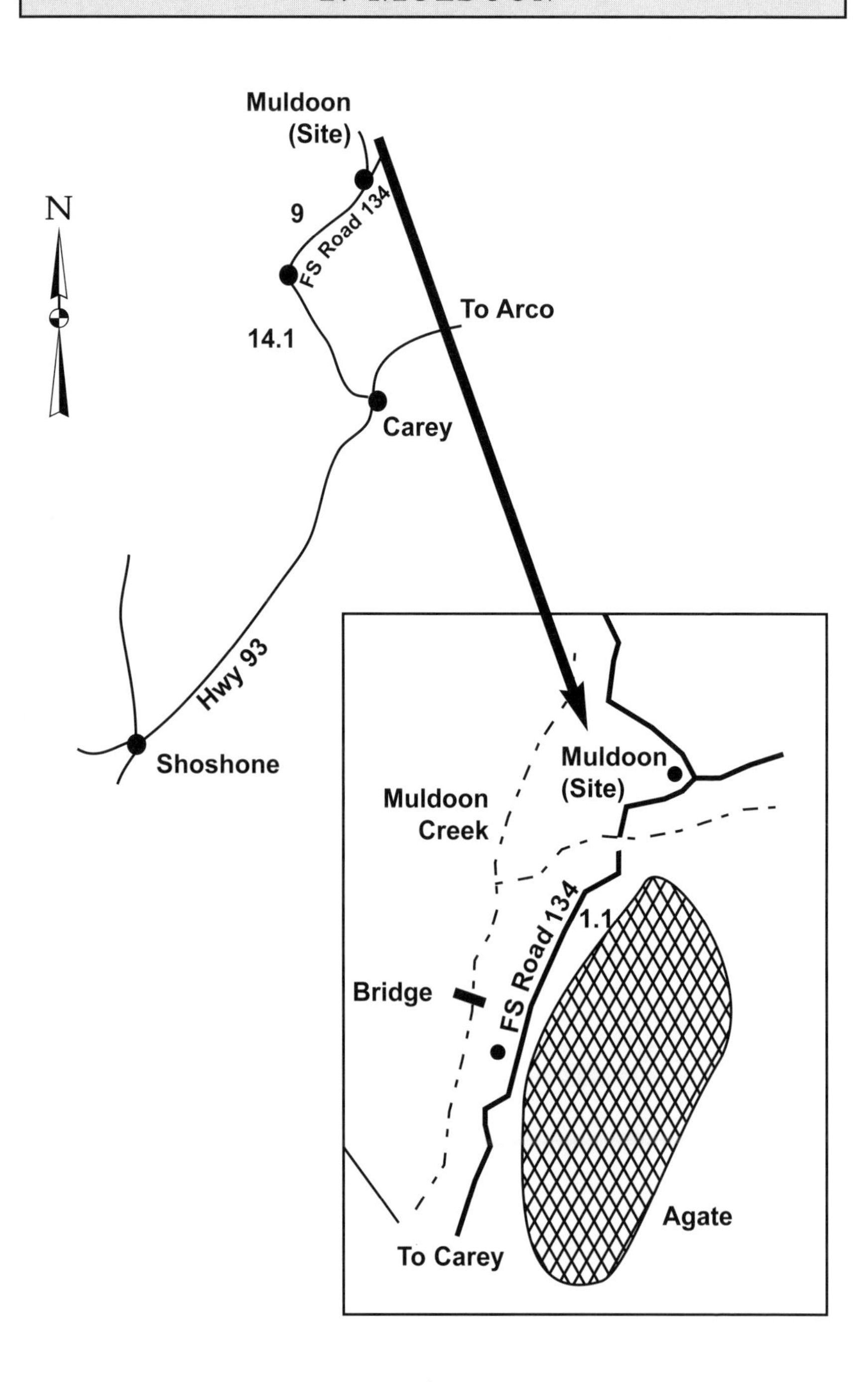

Muldoon
(Site)
N
9
FS Road 134
14.1
To Arco
Carey
Hwy 93
Shoshone
Muldoon
(Site)
Muldoon
Creek
FS Road 134
1.1
Bridge
Agate
To Carey

# 5. Cold Springs Creek

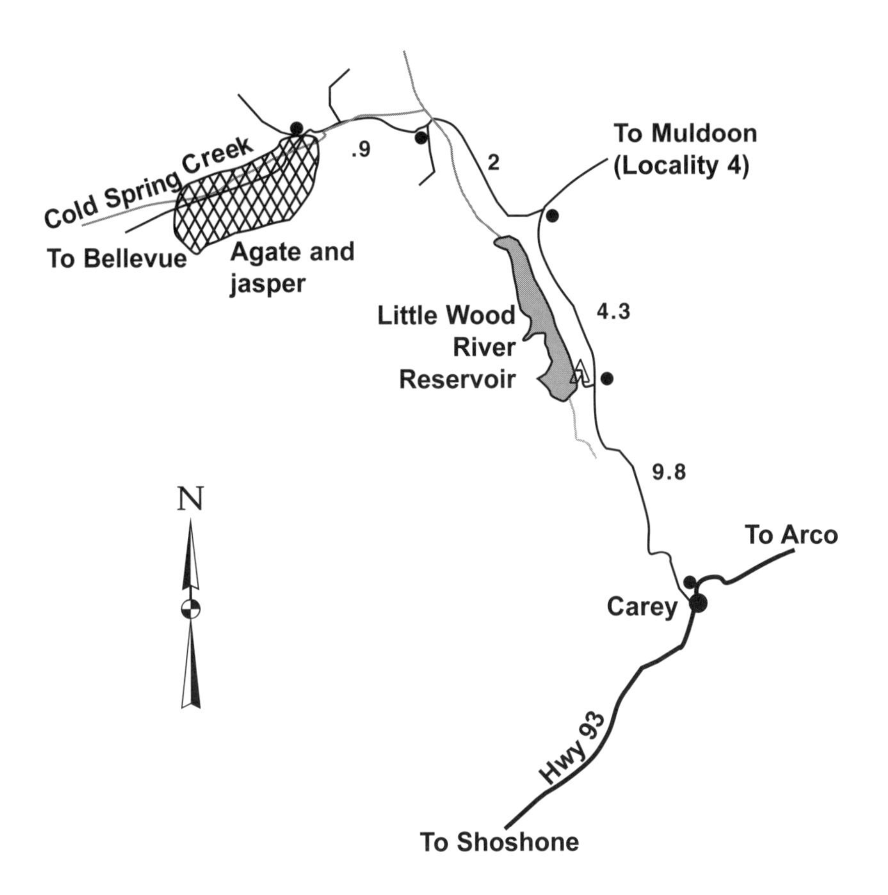

# 6A. & 6B. MACKAY

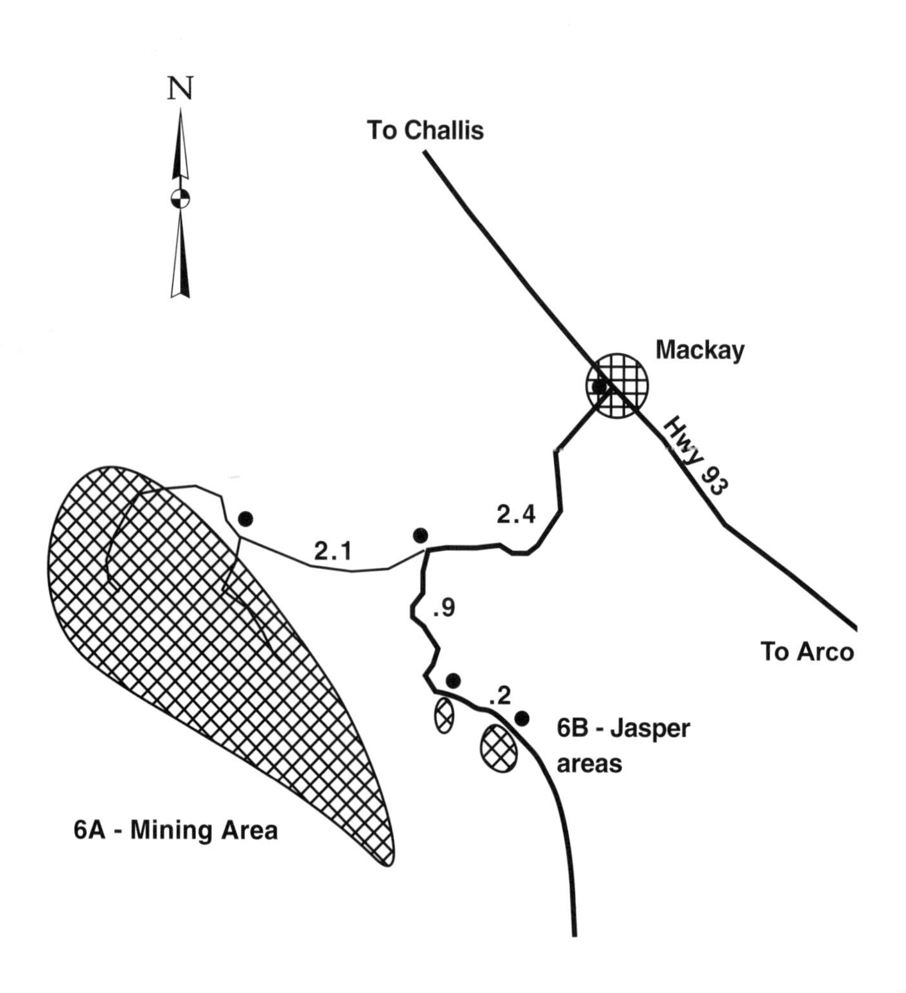
N
To Challis
Mackay
Hwy 93
2.4
2.1
.9
.2
To Arco
6B - Jasper areas
6A - Mining Area

# 7. Little Fall Creek

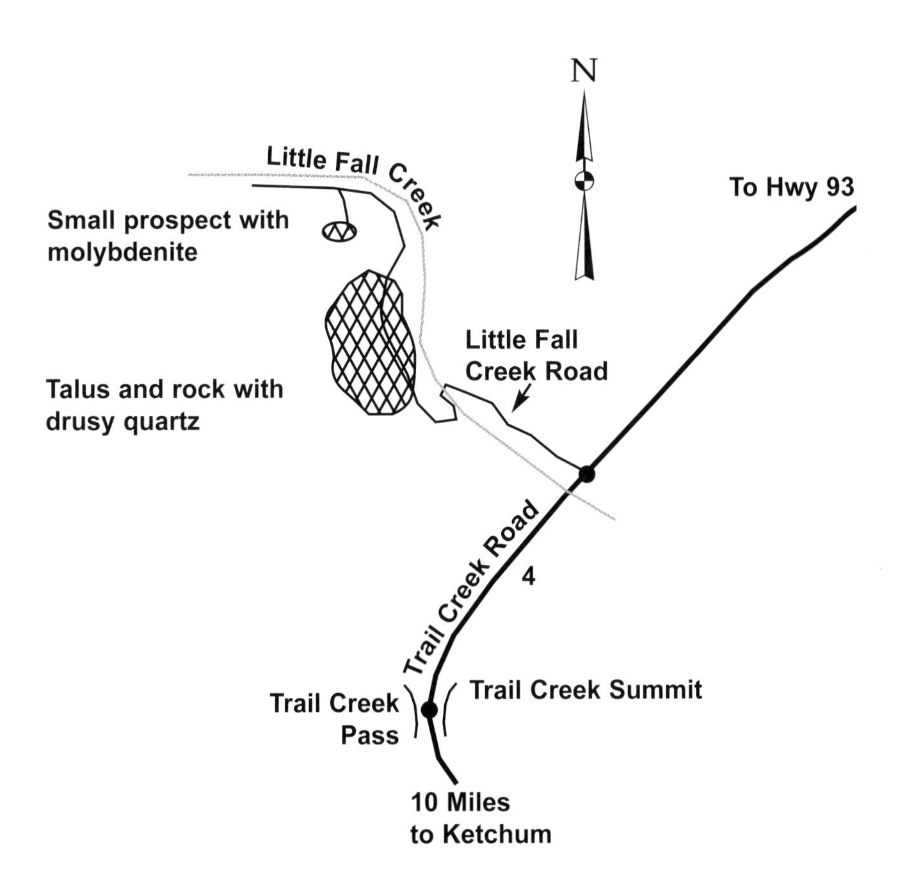

# 8. CHILLY CEMETERY

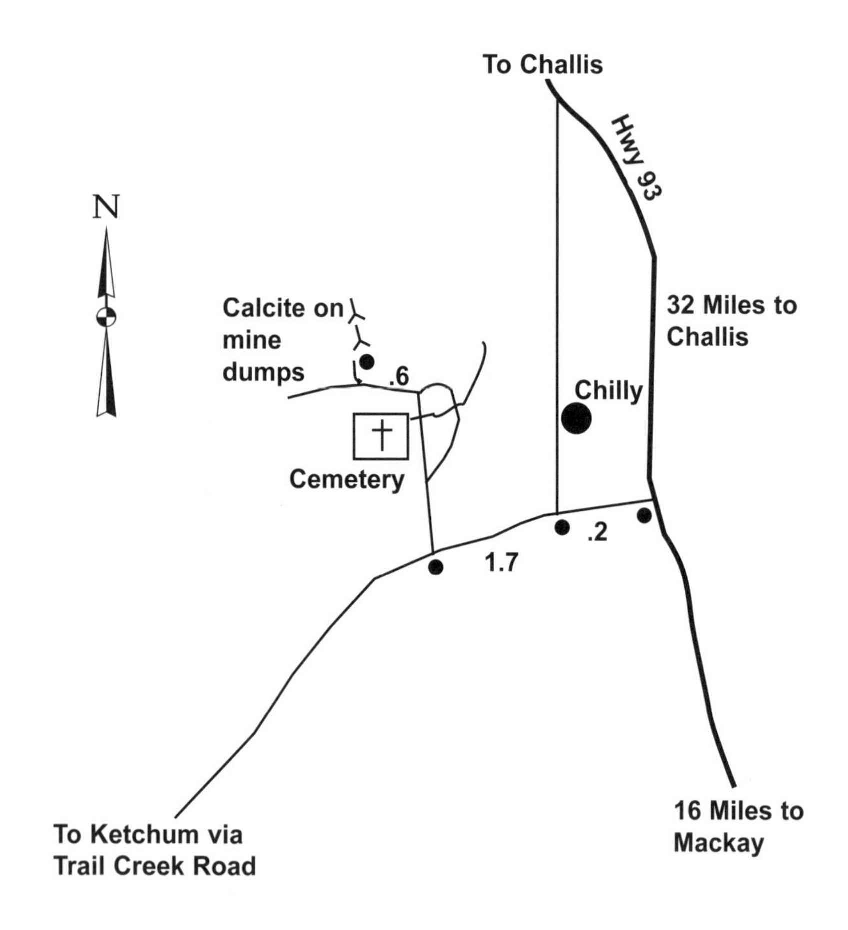

# 9. Lone Pine Creek

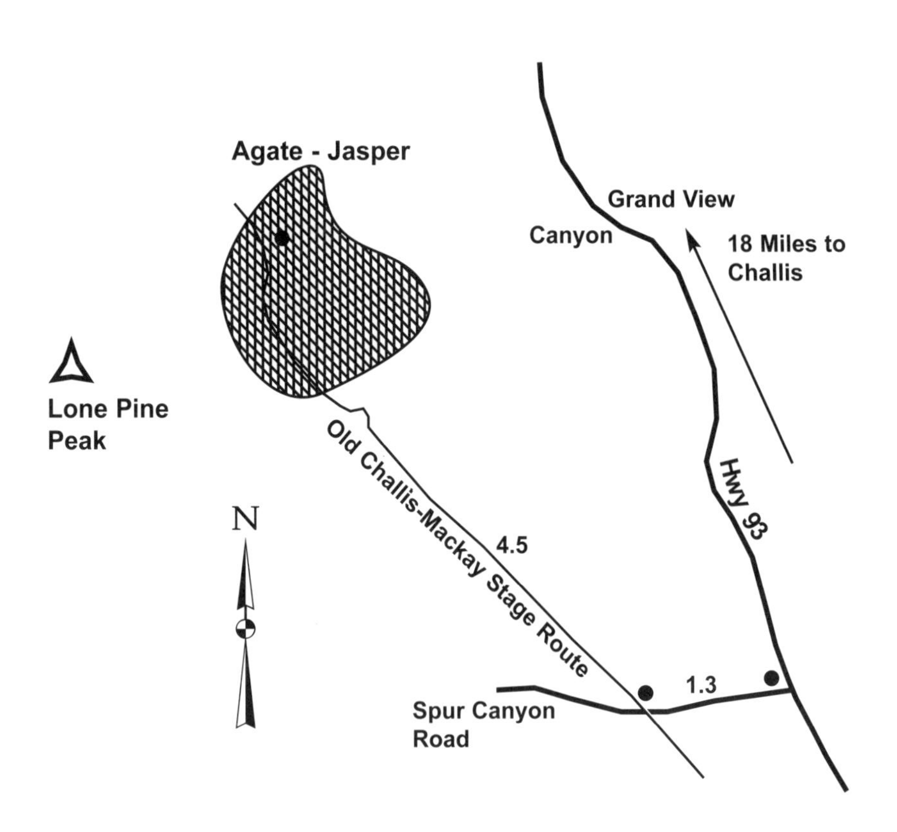

# 10. LIME CREEK – PASS CREEK

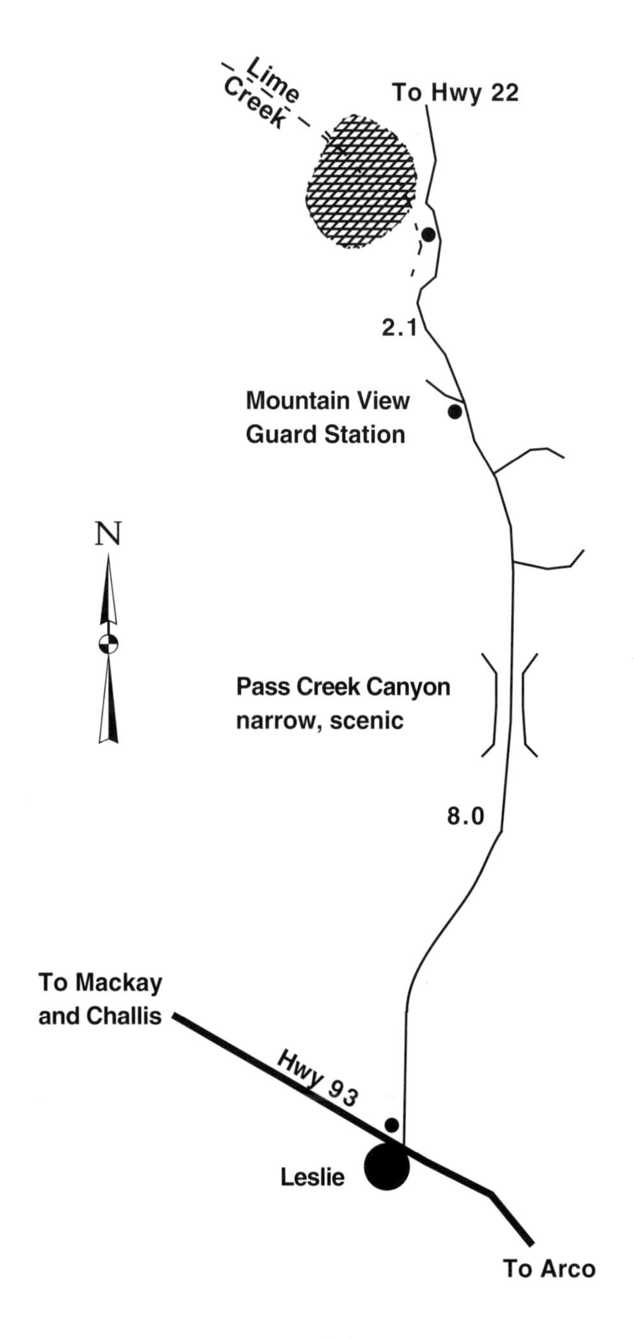

# 11A. & 11B. LIME CREEK

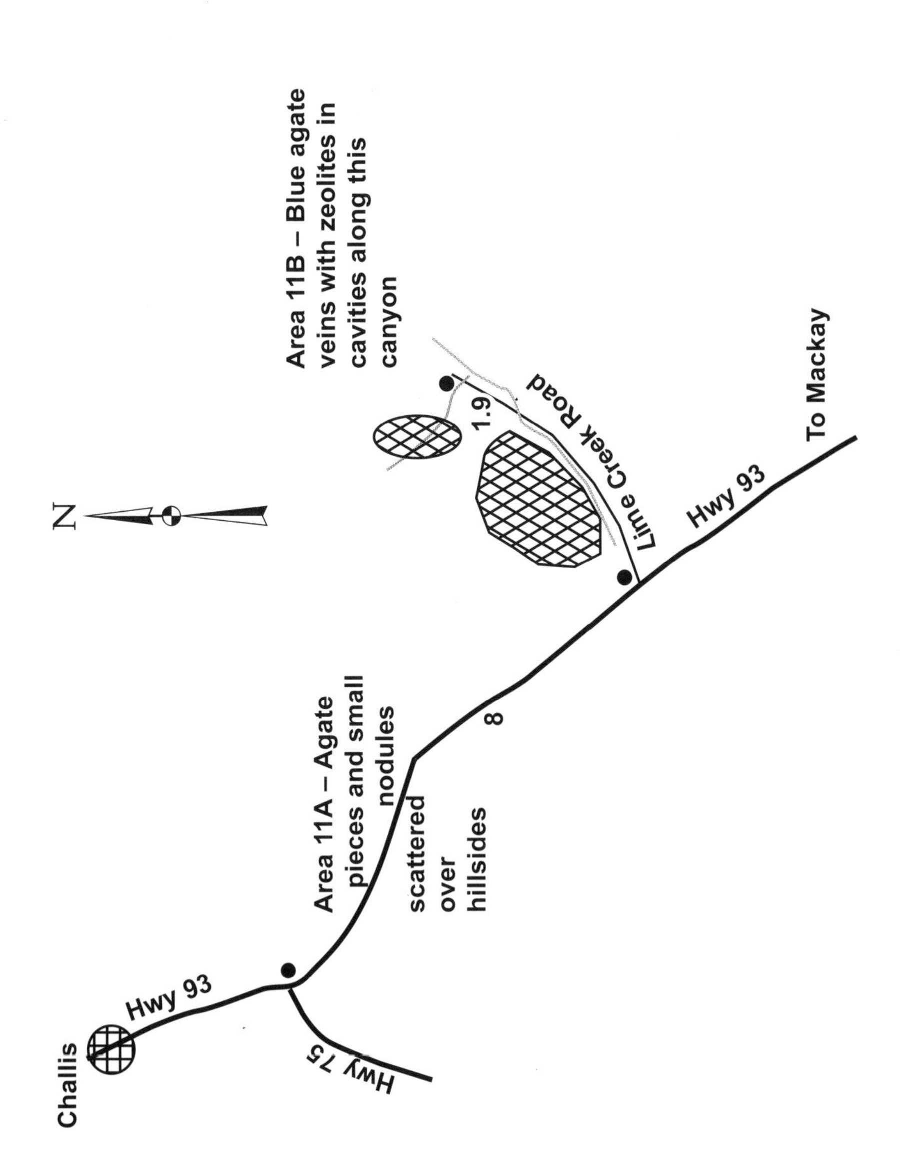

# 12. CINDER BUTTE & 13. SPENCER

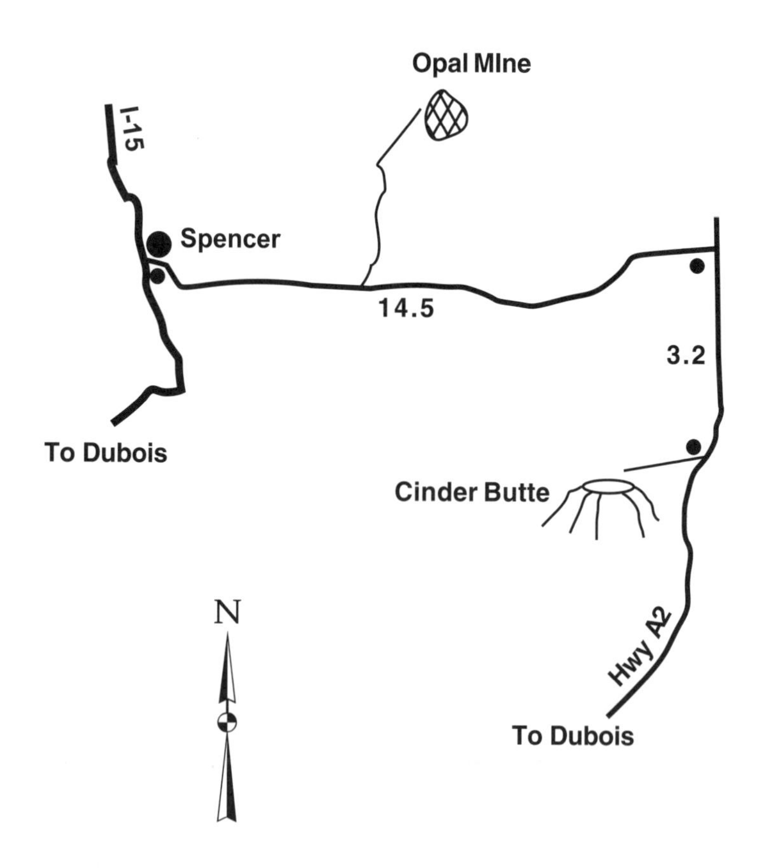

# 14A. & 14B. Road Creek

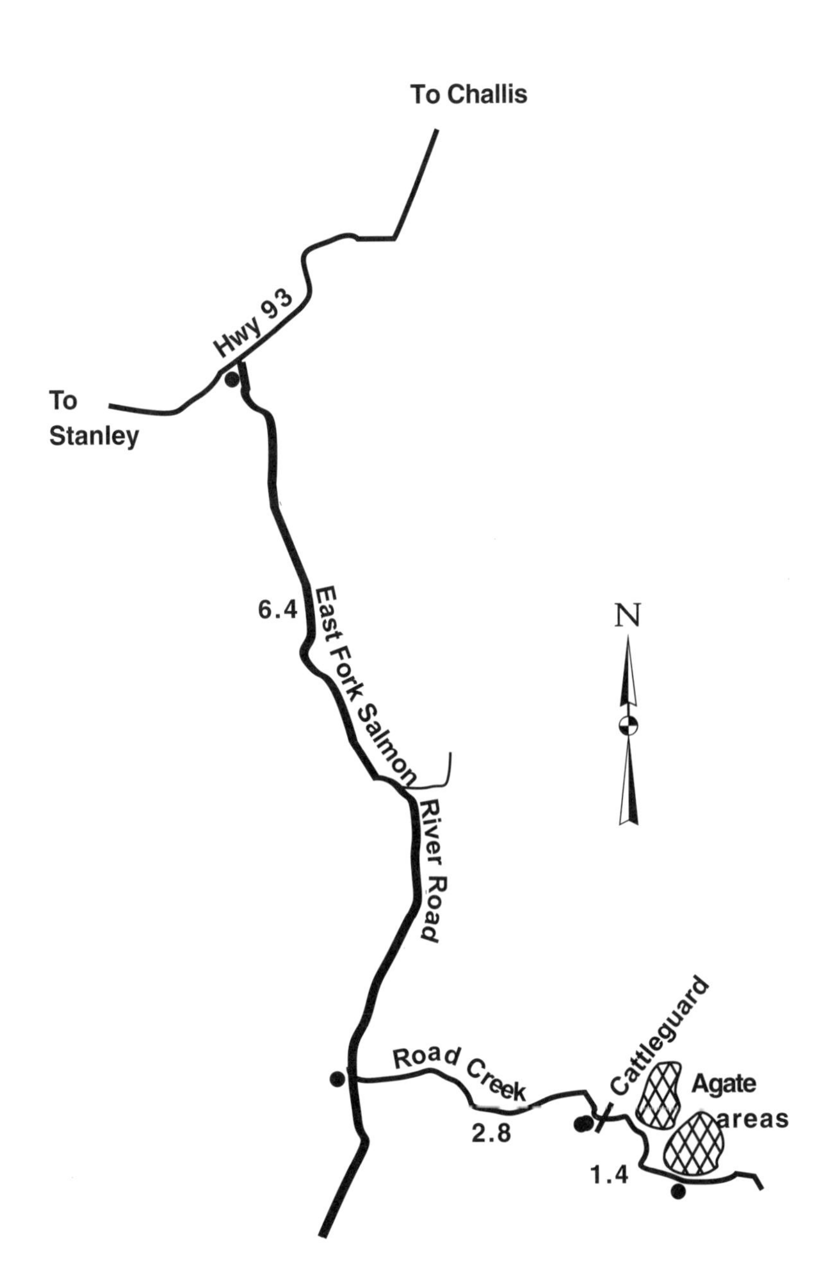

# 15. HERD CREEK

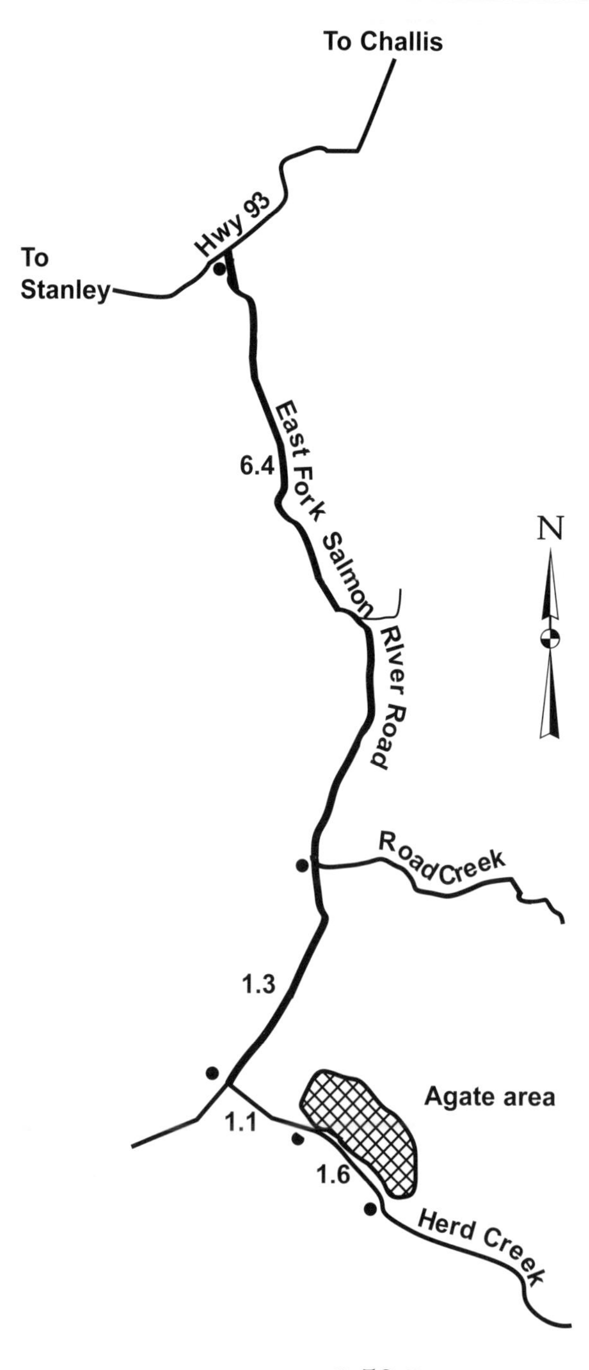

# 16. GARDEN CREEK

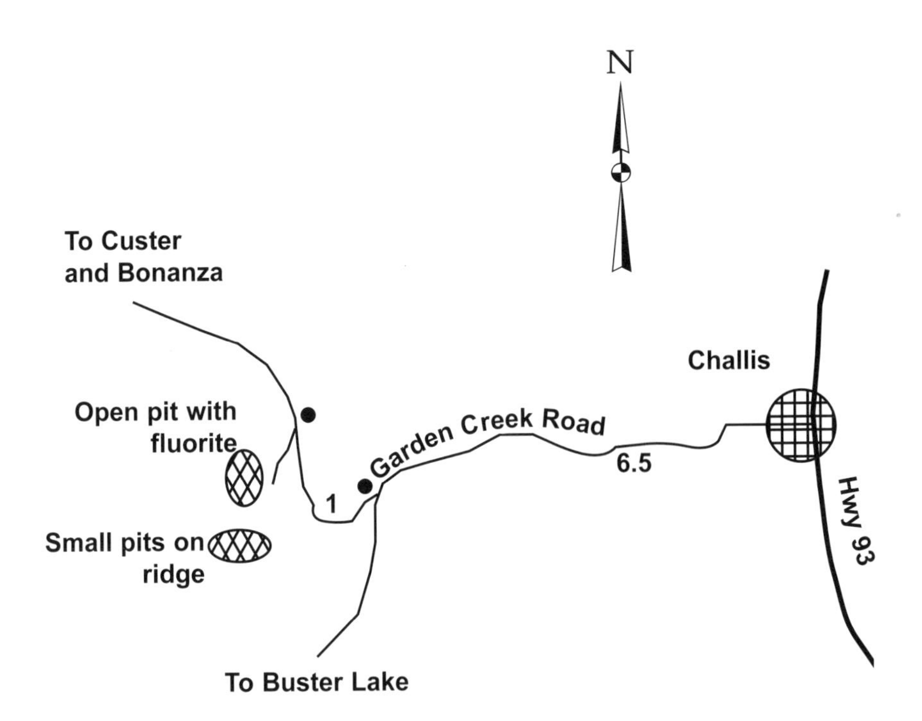

# 17. WILLIAMS CREEK

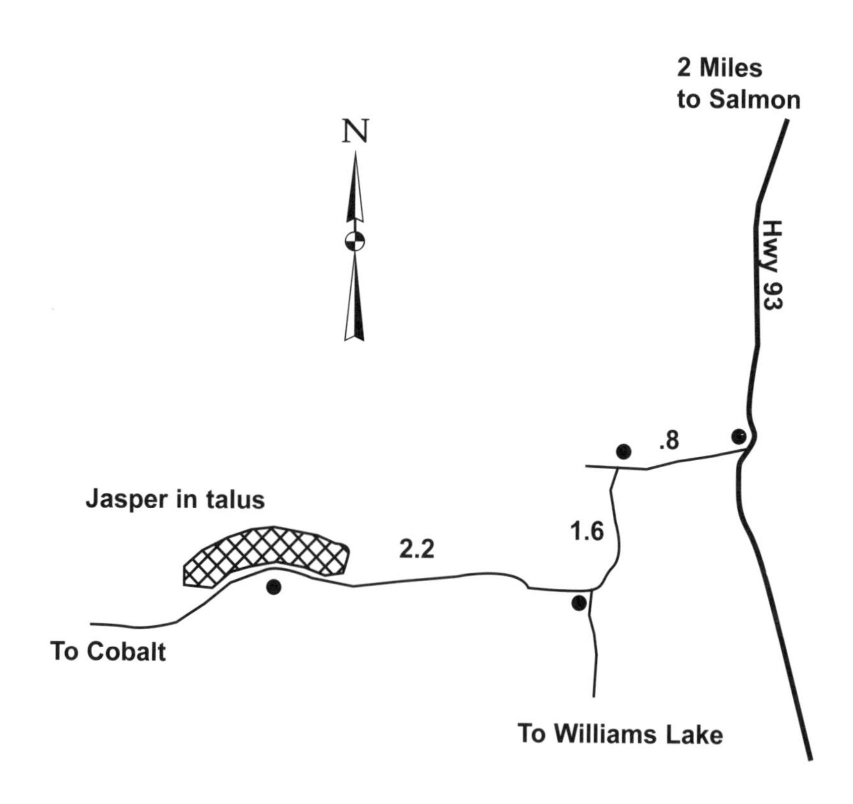

# 18. DISMAL SWAMP

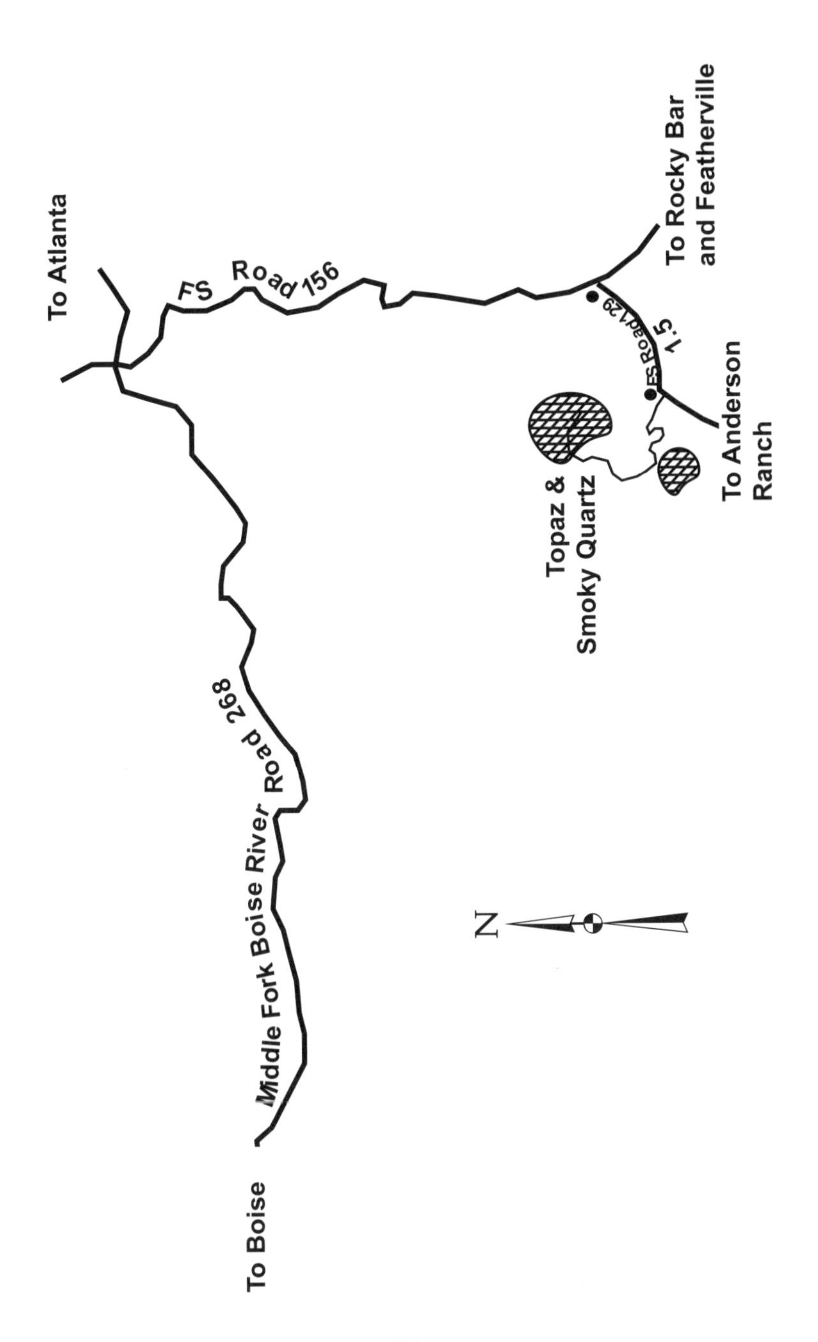

# 19. Lucky Peak Dam

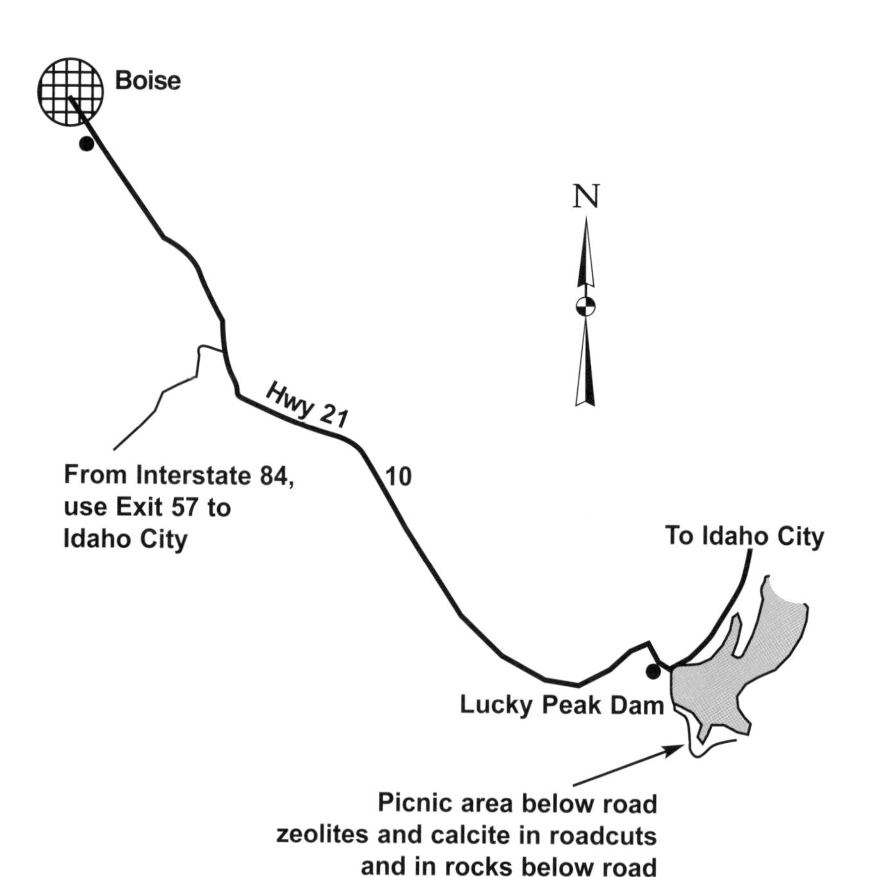

# 20. Sommercamp Road

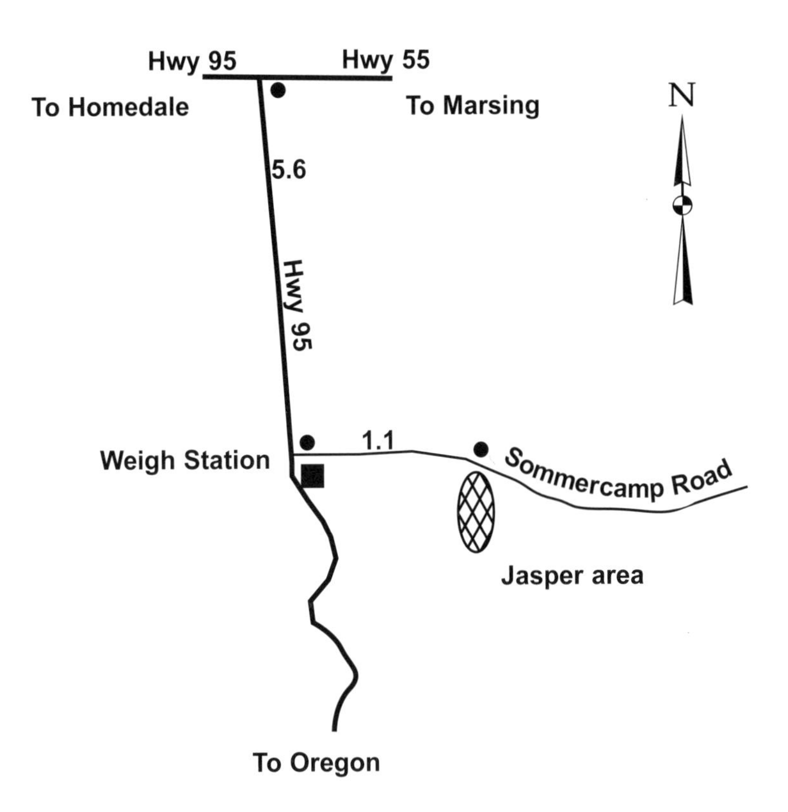

# 21. GRAVEYARD POINT

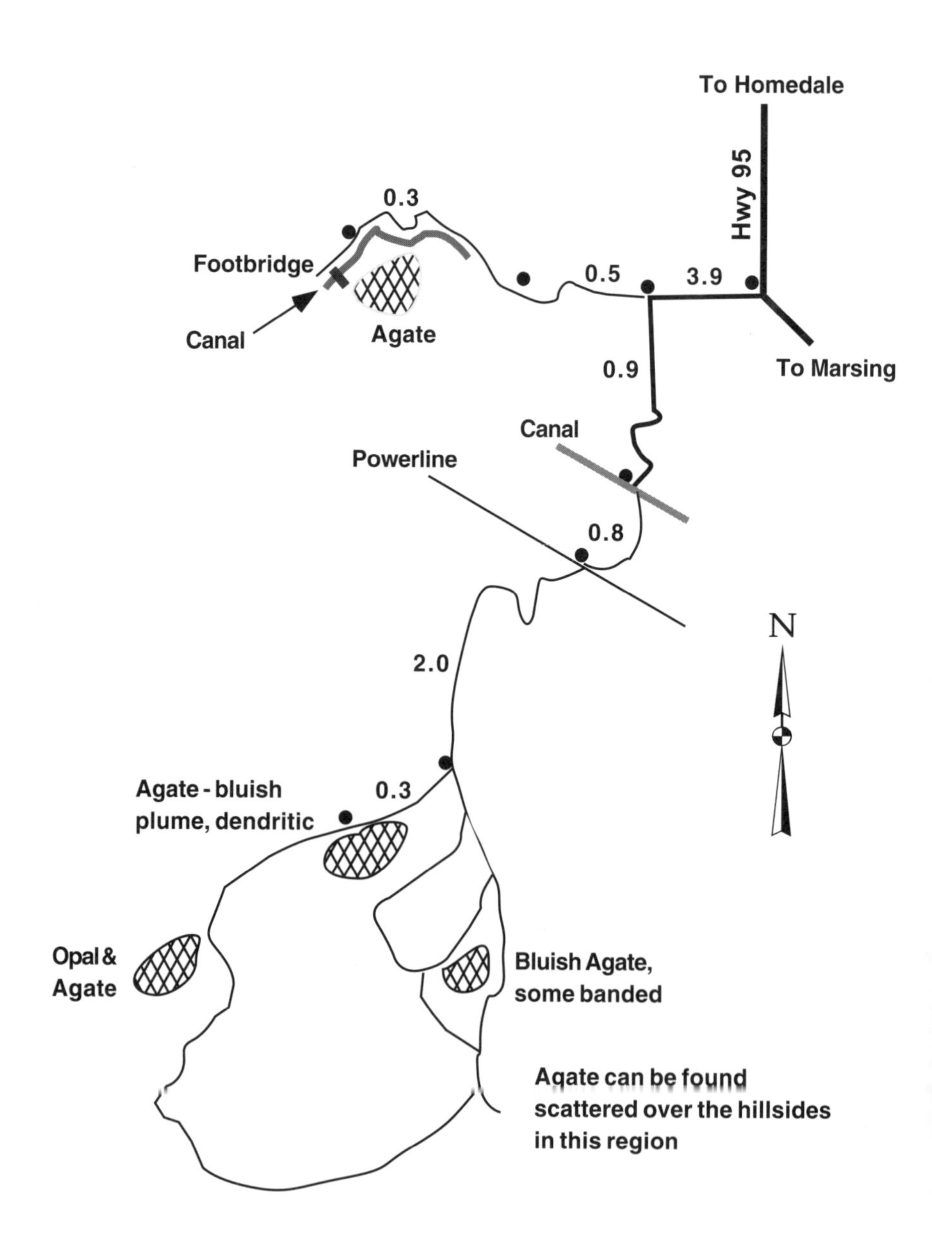

# 22. COAL MINE BASIN

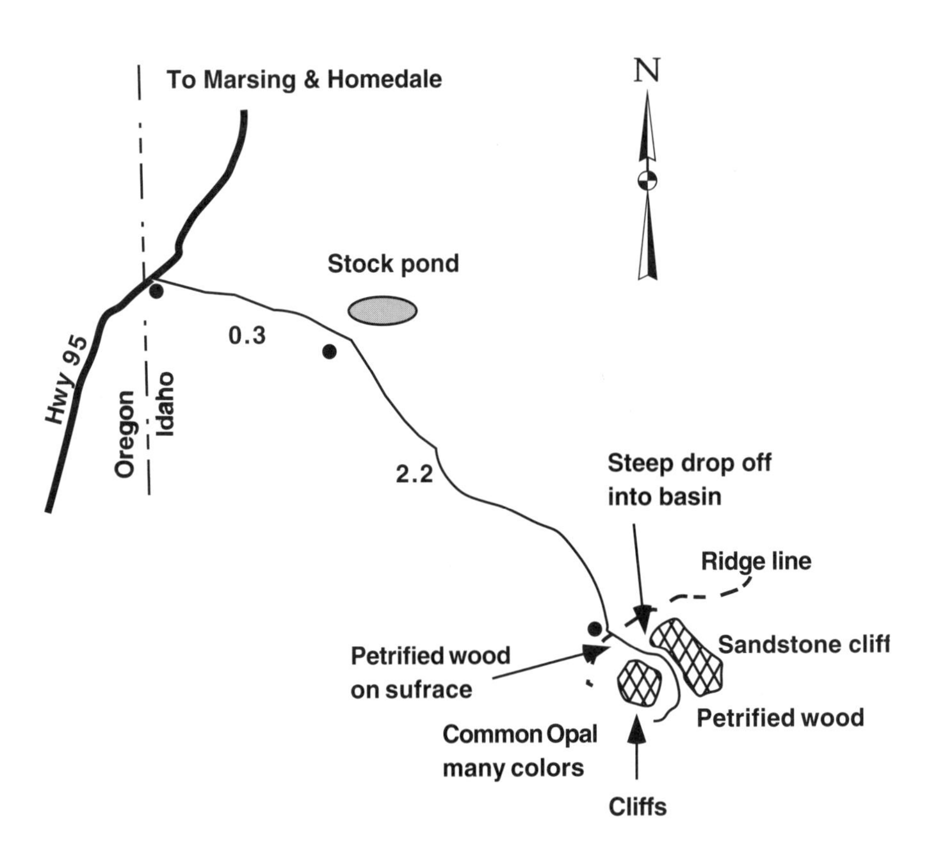

# 23A. & 23B. McBRIDE CREEK

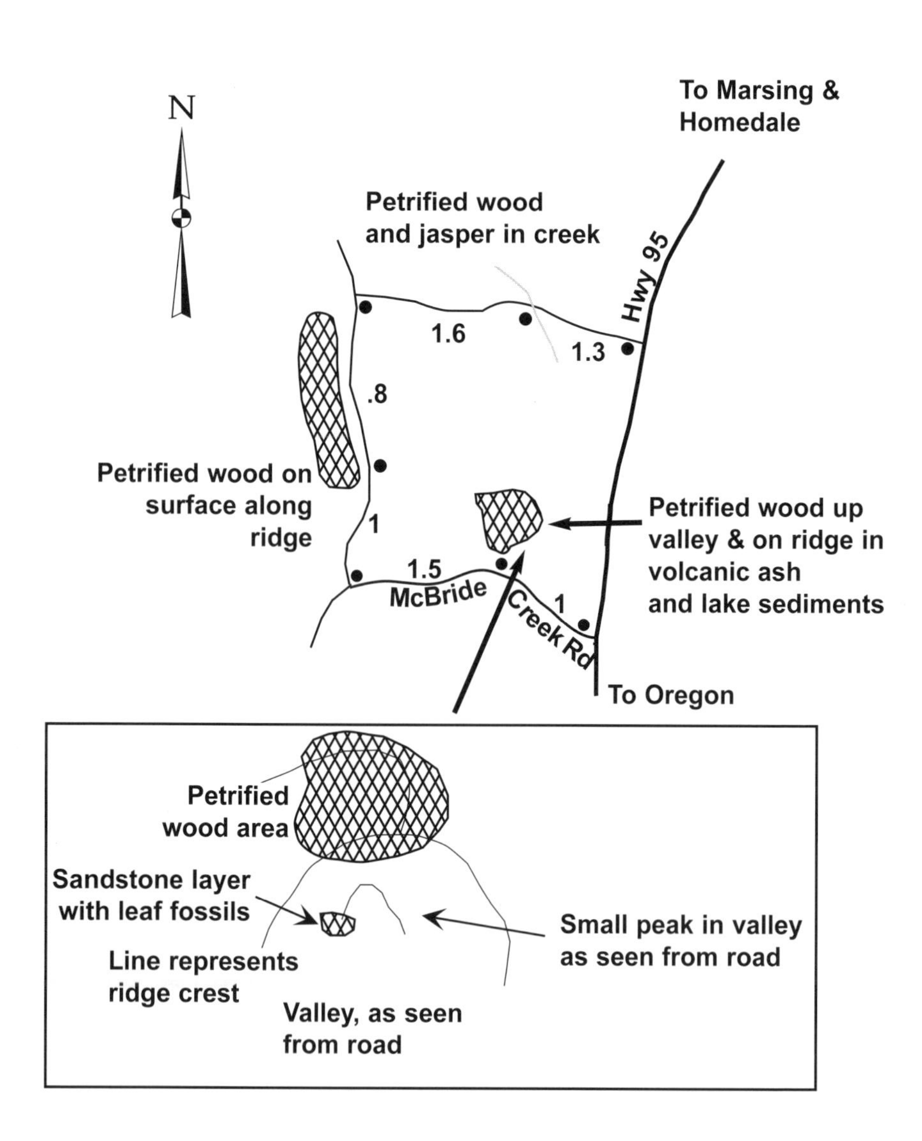

# 24. SOUTH MOUNTAIN

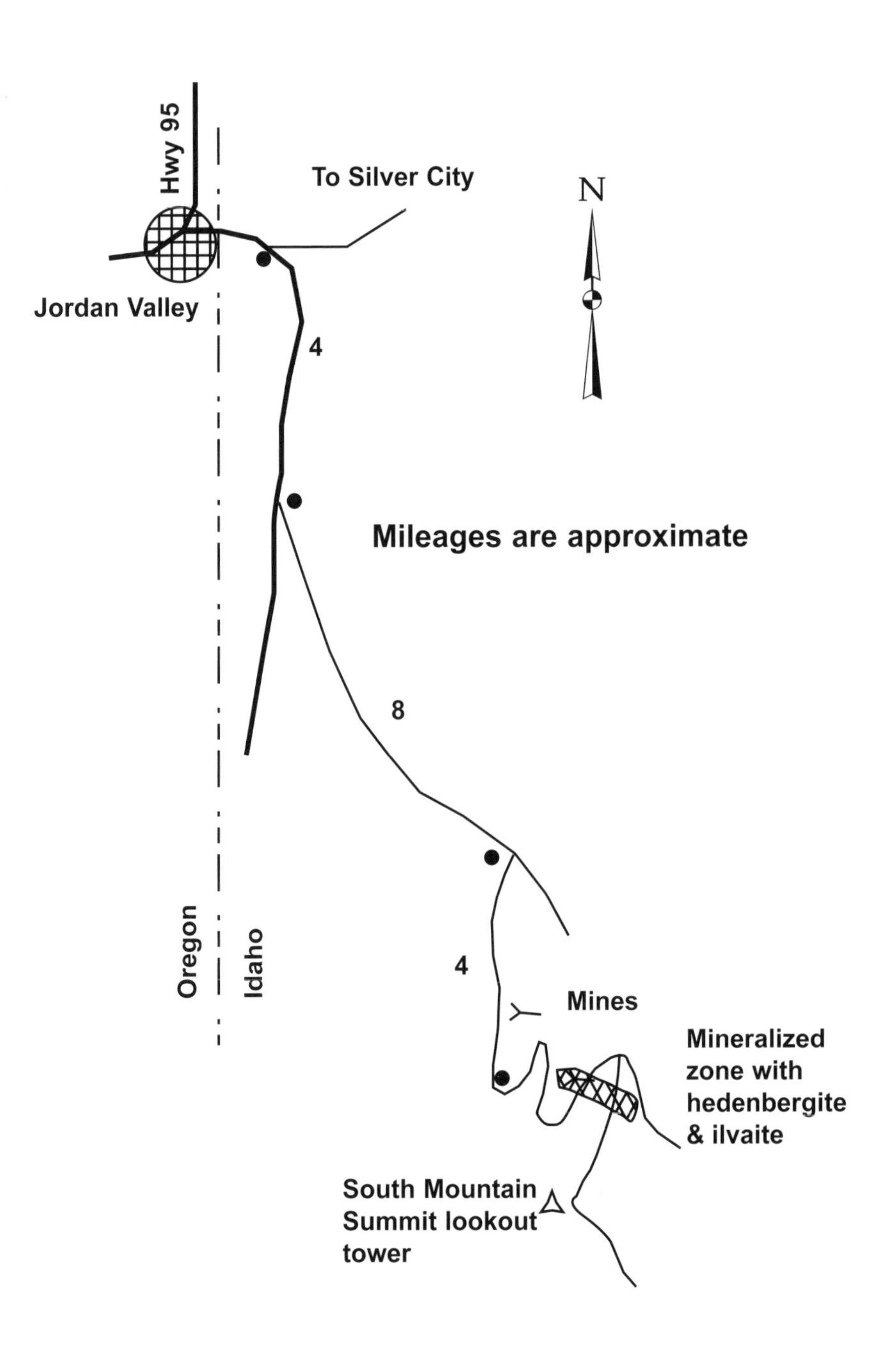

# 25. Beacon Hill

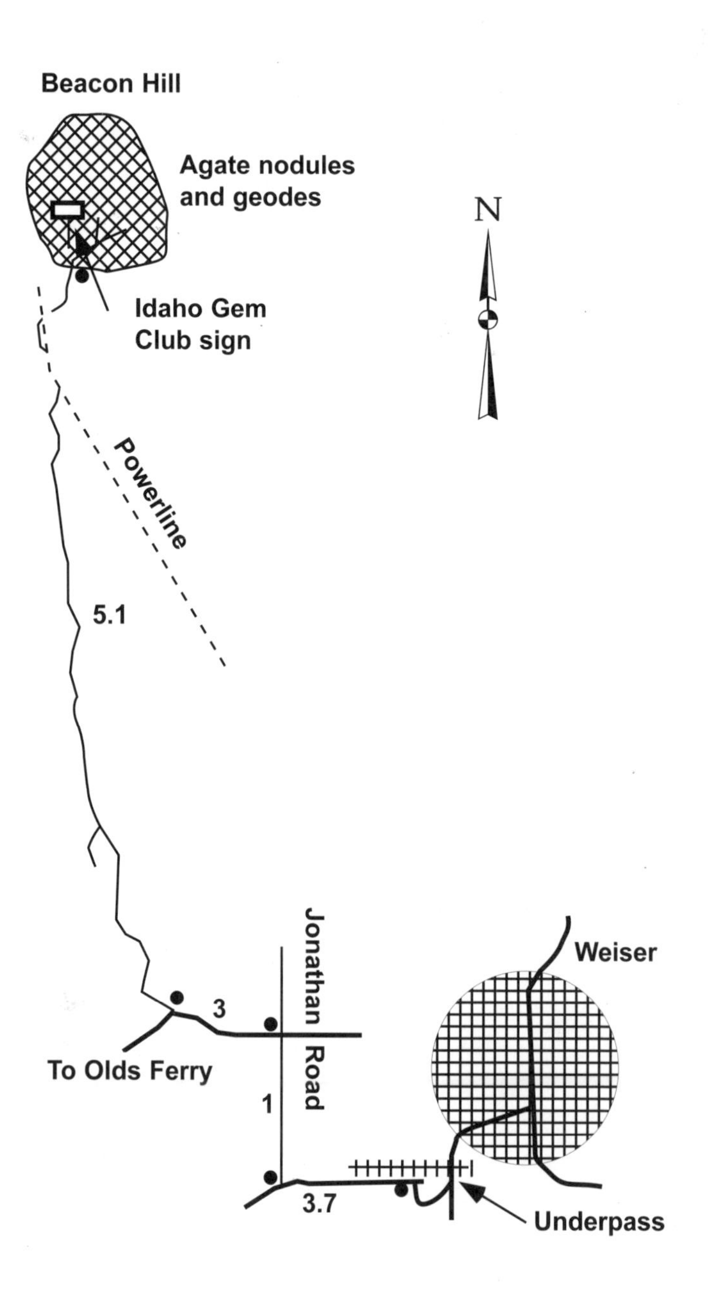

# 26. SEVEN DEVILS MINING DISTRICT

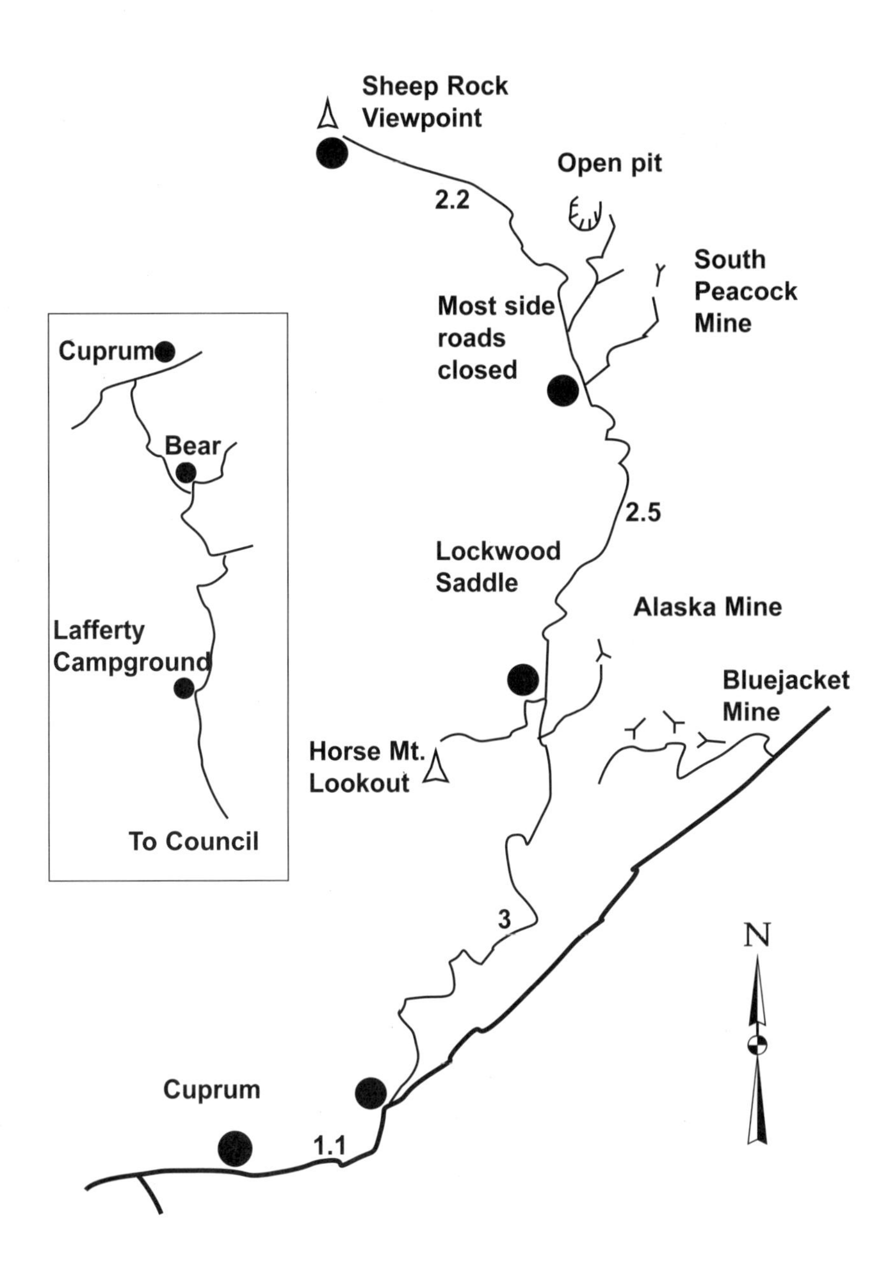

# 27. RUBY MEADOWS

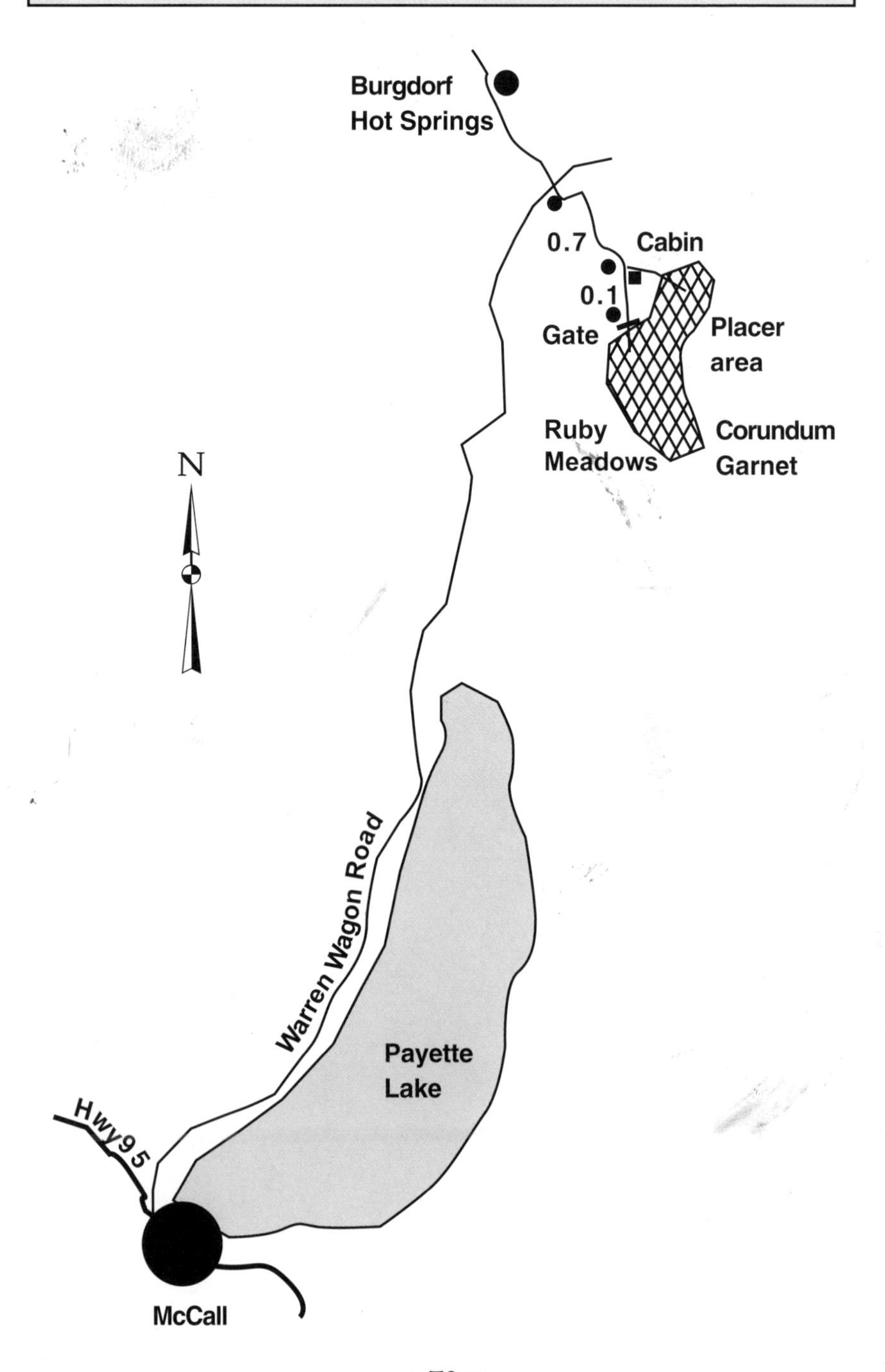

# 28. Pinehurst – Highway 95

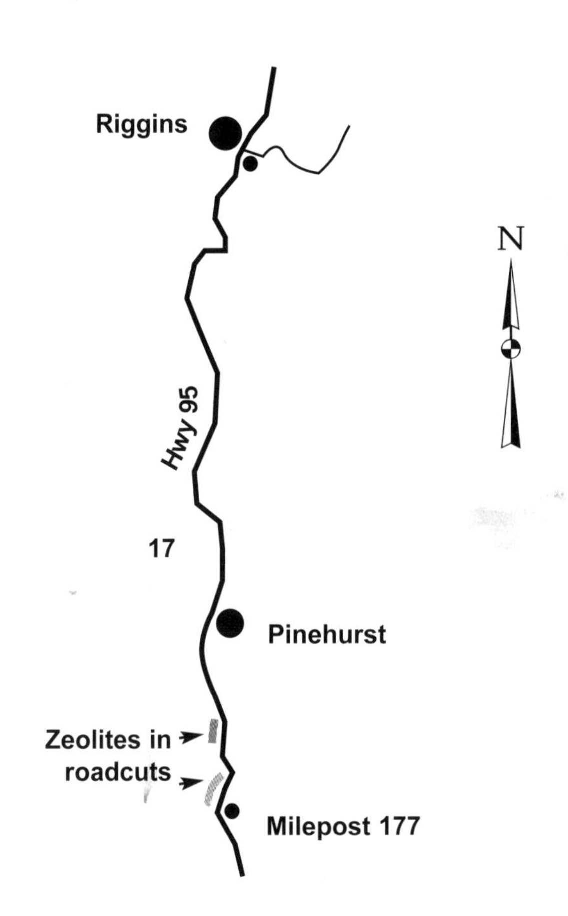

# 29. RUBY RAPIDS

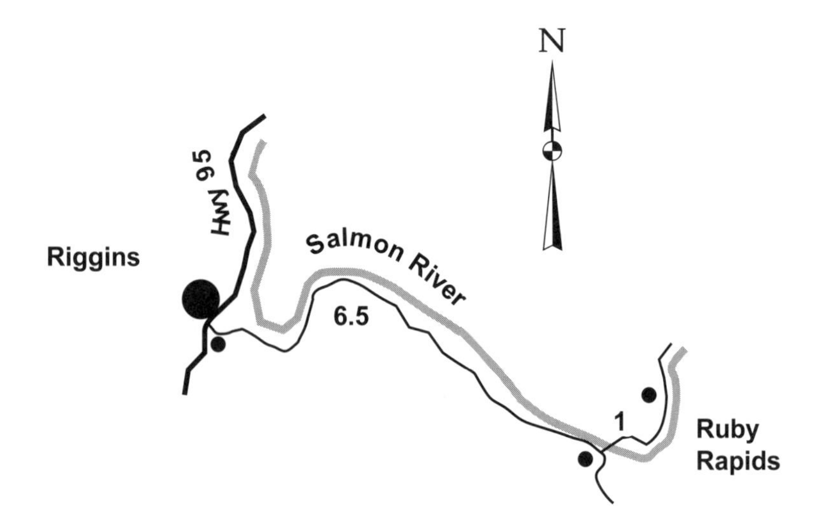

# 30. Slate Creek

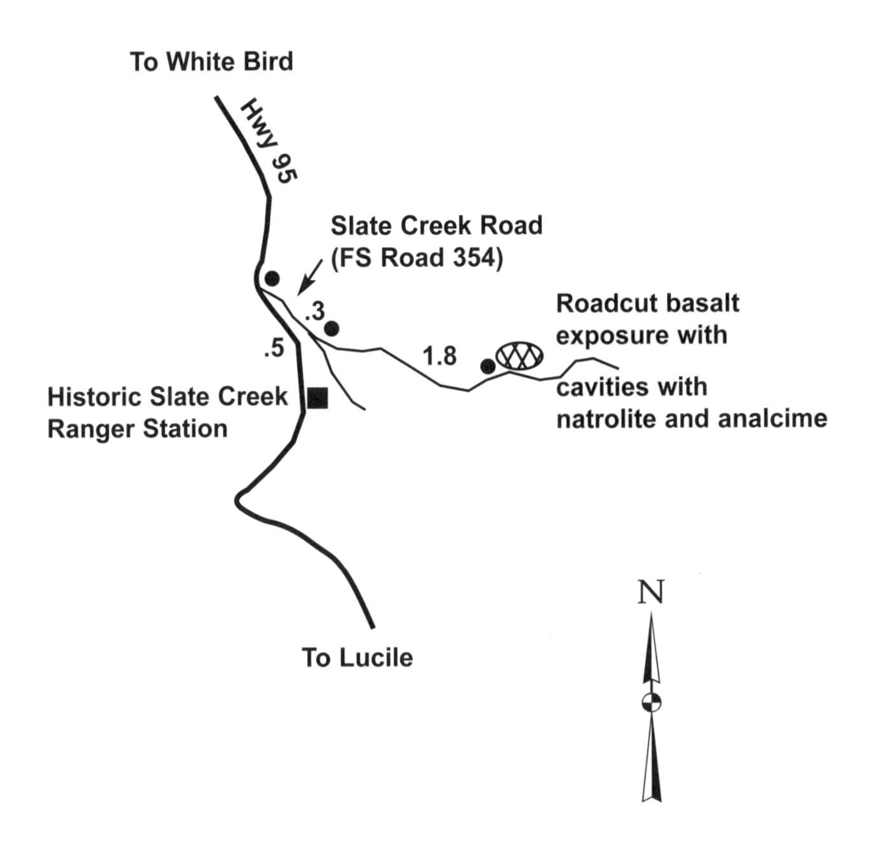

# 31. COTTONWOOD

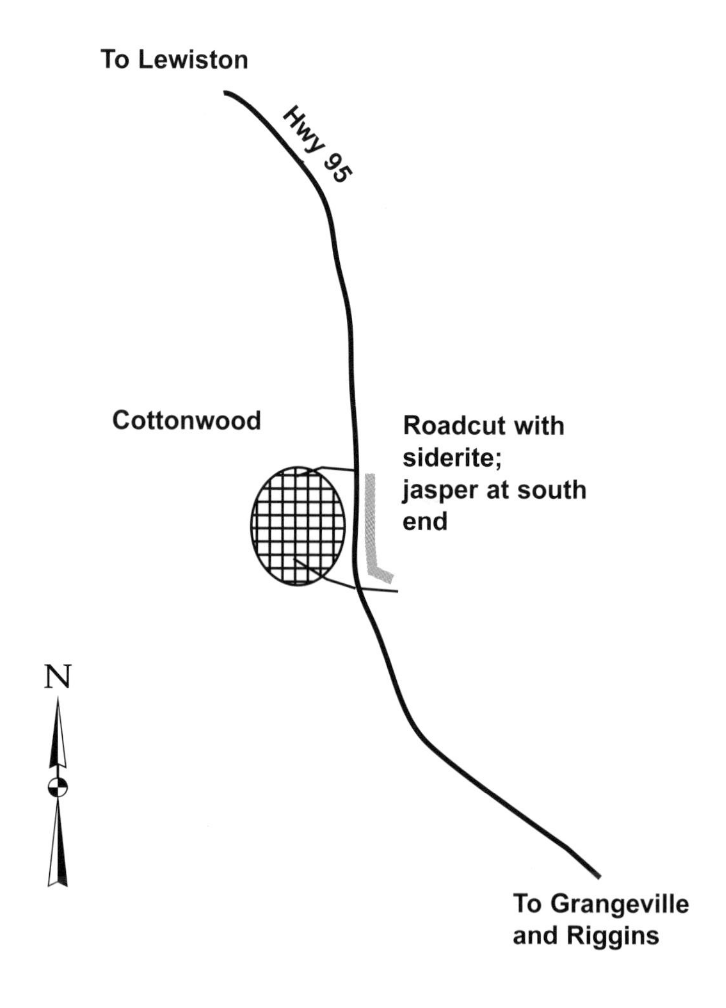

# 32. Mica Mountain

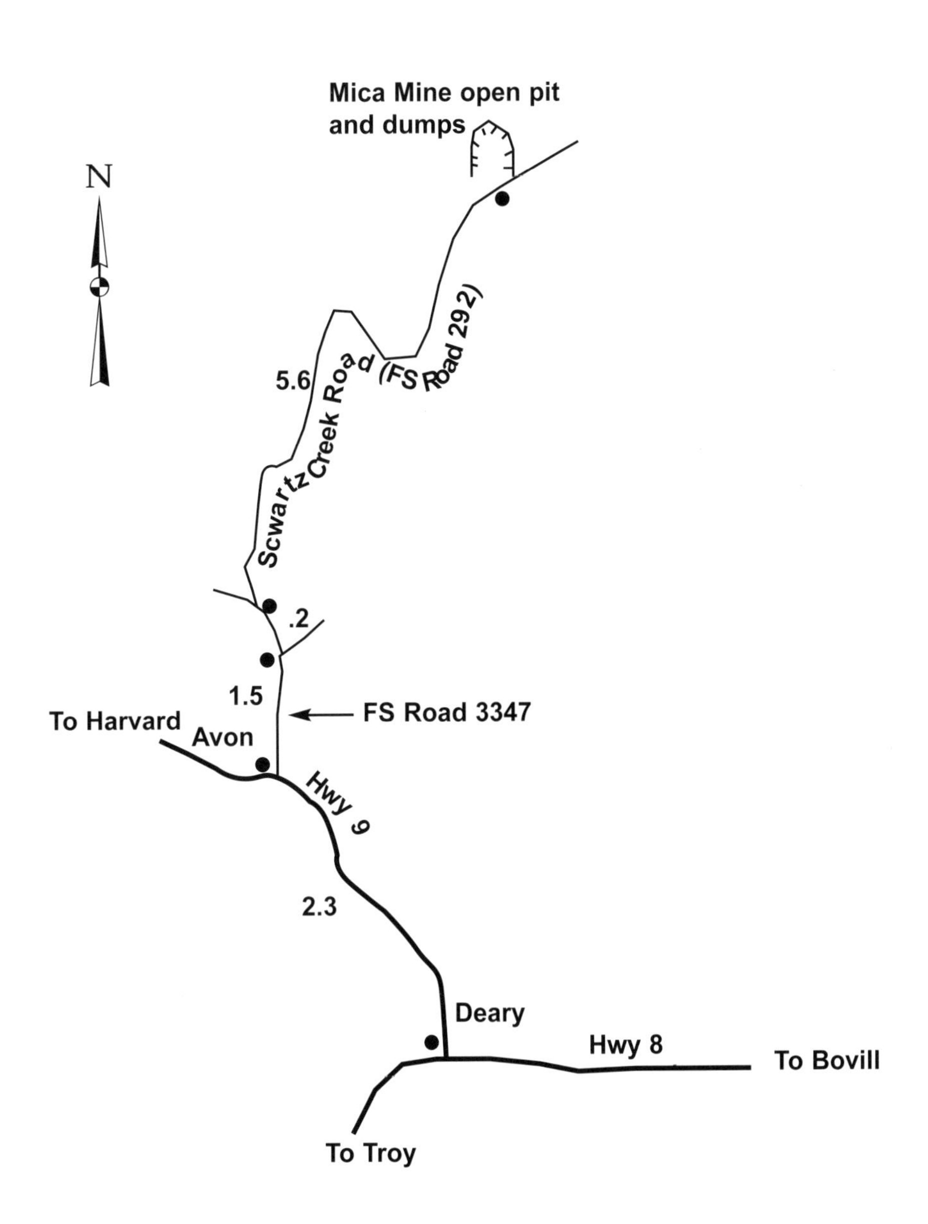

# 33. EMERALD CREEK

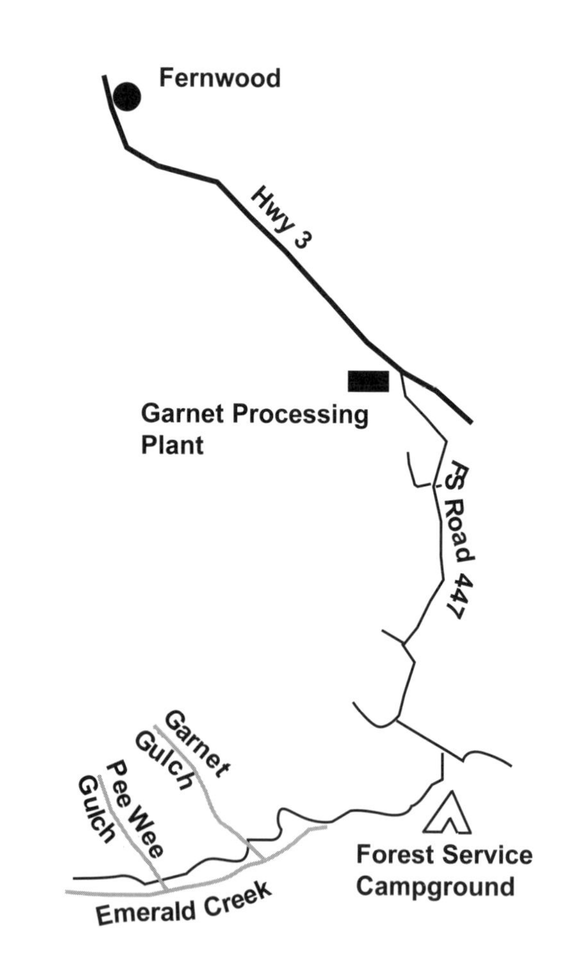

# 34. Freezeout Mountain

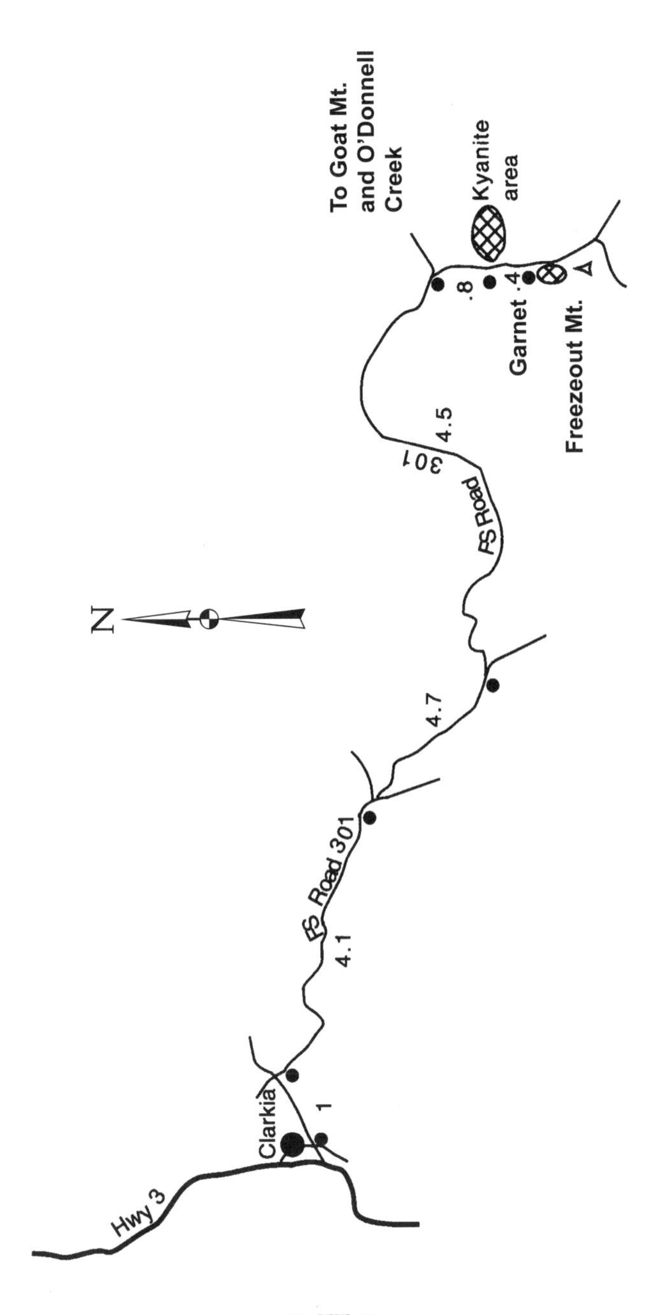

# 35. O'DONNELL CREEK

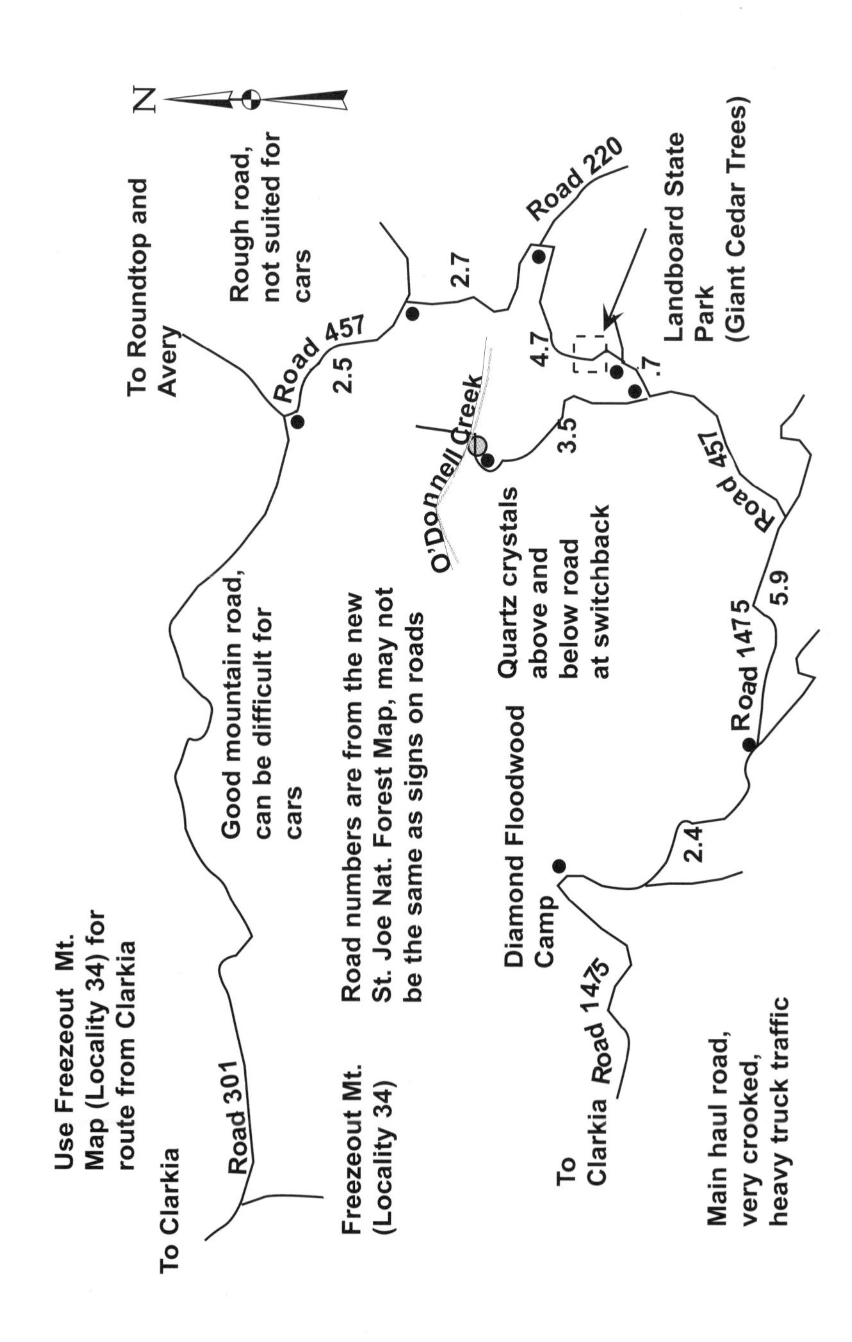

# 36. BATHTUB MOUNTAIN

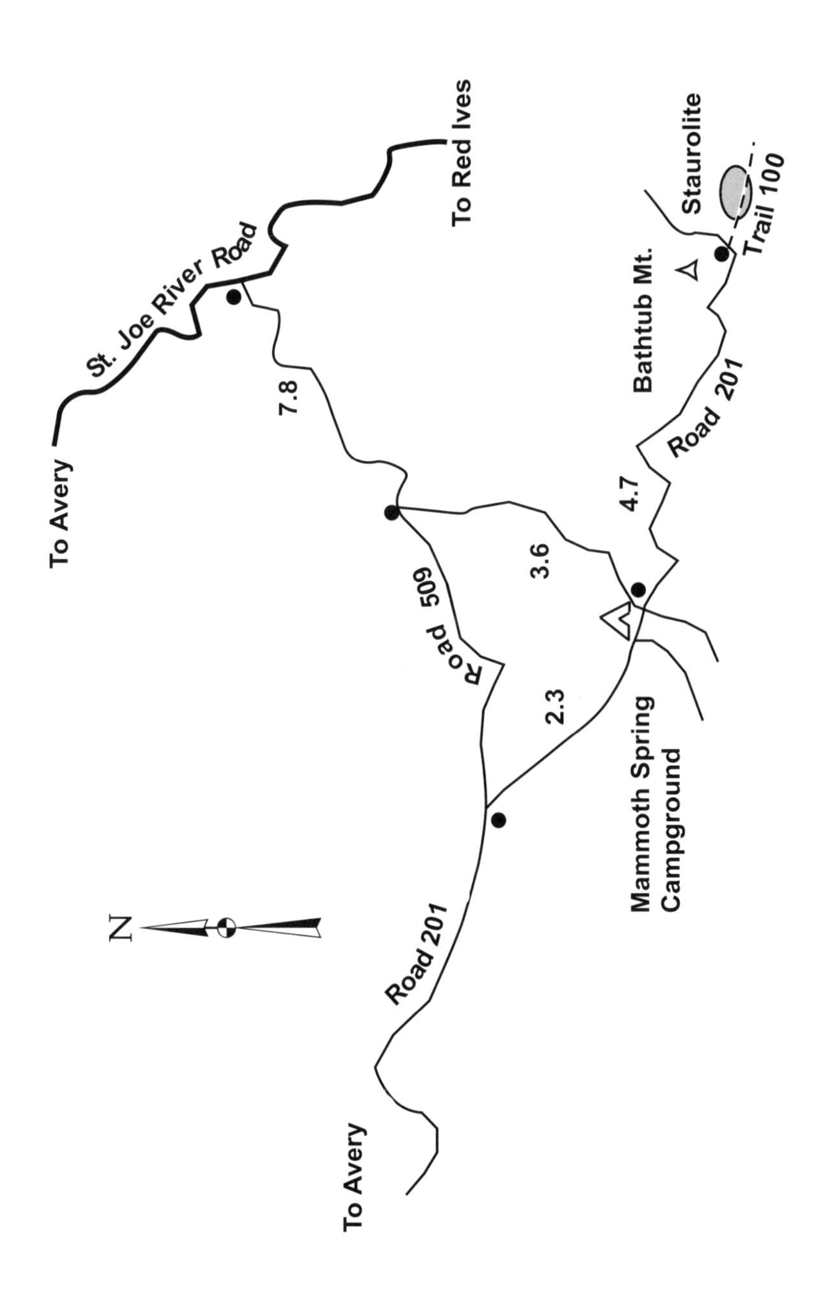

# NOTES